Cycads of Vietnam

With best wishes

Cycas pachypoda. **Plant showing the characteristic swollen 'foot' at the stem base, near Vinh Hoa village, Binh Thuan Province. Photo: Ken Hill.**

Cycads of Vietnam

written by

Roy Osborne
Ken D. Hill
Hiêp T. Nguyen
Loc Phan Ke

and published by

Roy Osborne and & Wynand van Eeden

ACKNOWLEDGEMENTS

Many persons and organisations were good enough to offer encouragement, advice and material assistance during the preparation of this volume.

Although most of the photographs were taken by the authors, additional material was kindly provided by Jeff Chemnick, Phillip Cribb, Jody Haynes and the Montgomery Botanical Center, Anders Lindström and Nong Nooch Tropical Garden, Damon Little, Todd Maerowitz, Joe Perner, Michael Perry, William Tang, Piet Vorster, Si Lin Yang and Ding Yue Wang. For their various and valued assistance in the 'Cycads of Vietnam' project, we are grateful to John Donaldson, Wes Field, Lesley Greenwood, Angela Osborne T/A Satooz.com, Lou & Tish Randall, Bart Schutzman and *The Cycad Newsletter*, Dennis Stevenson, William Tang, Piet Vorster, Stan and Jane Walkley and staff of Plantation 2000, Brisbane, Australia, staff of the National Herbarium of New South Wales, Royal Botanic Gardens, Sydney, Australia, and staff of Tien Wah Press (Pte) Ltd, Singapore. Permission to use illustrations appearing elsewhere was kindly granted by the Publications Department of the New York Botanical Garden, the Royal Botanic Gardens and Domain Trust of Sydney, Australia, and the editor of *Encephalartos*, Journal of the Cycad Society of South Africa. Finally, we acknowledge the cordial friendship of all participants on the '2002 Vietnam Cycad Expedition' (see Vorster 2002, 2003a & b), which inspired the production of this volume.

Additional acknowledgement: The authors wish to acknowledge the kind cooperation of the Institute of Ecology and Biological Resources (IEBR), Vietnam Academy of Science and Technology, which has been a major contributor to cycad studies in Vietnam. Without their ongoing support, the cycad studies in Vietnam would not have been carried out and this text would not have been published.

National Library of Australia
Cataloguing-in-Publication data:

Osborne, Roy, 1944-.
Cycads of Vietnam.

Bibliography.
Includes index.
ISBN 978-0 646-46445-9

1. Cycads - Vietnam. I. Title.
585.909597

Design and layout by	Wynand van Eeden
Published by	Roy Osborne, Brisbane, Australia, & Wynand van Eeden, Cape Town, South Africa
Printed by	Tien Wah Press (Pte) Ltd, Singapore
Distributed by	Satooz (http://www.satooz.com)

CONTENTS

Cycas condaoensis. Herbarium specimen, _L. Pierre 1722_, collected in September 1877, the Paris isotype to the Kew holotype for the species. Photo: Ken Hill.

FOREWORD

John S. Donaldson, Ph.D., F.R.E.S.

Chair, Cycad Specialist Group, Species Survival Commission,

World Conservation Union (IUCN).

Vietnam has become recognised as a country of enormous biological diversity, both in its fauna and flora. Surveys over the last two decades have revealed an abundance of cycads—fascinating plants now known to occupy a key position in plant evolution. All the cycads in Vietnam belong in the genus *Cycas*, which is arguably the oldest, the largest, and the most widespread of the 11 cycad genera. It is also the genus that has contributed most to the large and growing trade in ornamental cycads. Vietnam boasts 27 different species of this genus and ranks with South Africa, Australia, and Mexico as a centre of endemism for cycads.

Vietnam's cycads, many of which have been named only in the last 10–15 years, fall within three of the *Cycas* sections and occupy a variety of specialised habitats. Some Vietnamese cycads are obligate beach-dwellers (*Cycas litoralis*), some live only in the northern mountains (*C. collina*), others occur on small offshore islands (*C. condaoensis*), and some species even grow in tiny crevices on near-vertical limestone cliffs (*C. ferruginea*). These unique plants form an integral part of Vietnam's natural heritage and contribute to the spectacular natural beauty of areas such as Ha Long Bay (*C. tropophylla*) and Ba Be National Park (*C. brachycantha*).

Cycads are often described as 'palm-like', but this generalisation masks the amazing diversity of growth forms found among cycads. Vietnam's cycads include species that have tree-like growth forms, reaching 12 m in height (*C. pectinata*), while others are bonsai-like dwarfs (*C. hoabinhensis*). Species such as *C. brachypoda* have strangely swollen 'elephant's foot' trunks, while other species have unusual pinnate leaves reaching 4.5 m in length (*C. dolichophylla*) or multiple leaf bifurcations (*C. multipinnata*), revealing their ongoing adaptation to different environments.

In '*Cycads of Vietnam*', each of the 27 species is described in detail with line drawings, maps and numerous photographs illustrating points of interest. Additional notes deal generally with Vietnam's history, geography, climate and biology. An informative chapter is devoted to a summary of important aspects of cycad biology.

'*Cycads of Vietnam*' is the product of four talented and knowledgeable cycad authorities and they need to be commended for this contribution to the cycad literature. Roy Osborne, founder and first President of the Cycad Society of South Africa, is a well-known and highly respected cycad writer. His Australian colleague Ken Hill is the world's acknowledged expert on the genus *Cycas*. Hiệp T. Nguyen and Loc Phan Ke are two of Vietnam's leading botanists. All of them are past or present members of the IUCN/SSC Cycad Specialist Group whose collective knowledge of the cycads of Vietnam is immense.

I have no doubt that this book will satisfy many different interests, from cycad hobbyists and nurserymen to researchers, students, conservation officials and those with a general interest in plant evolution and botany. I trust that discovering more about the cycads of Vietnam will mean that we all have a greater appreciation for the splendour, diversity, and fragility of these magnificent plants.

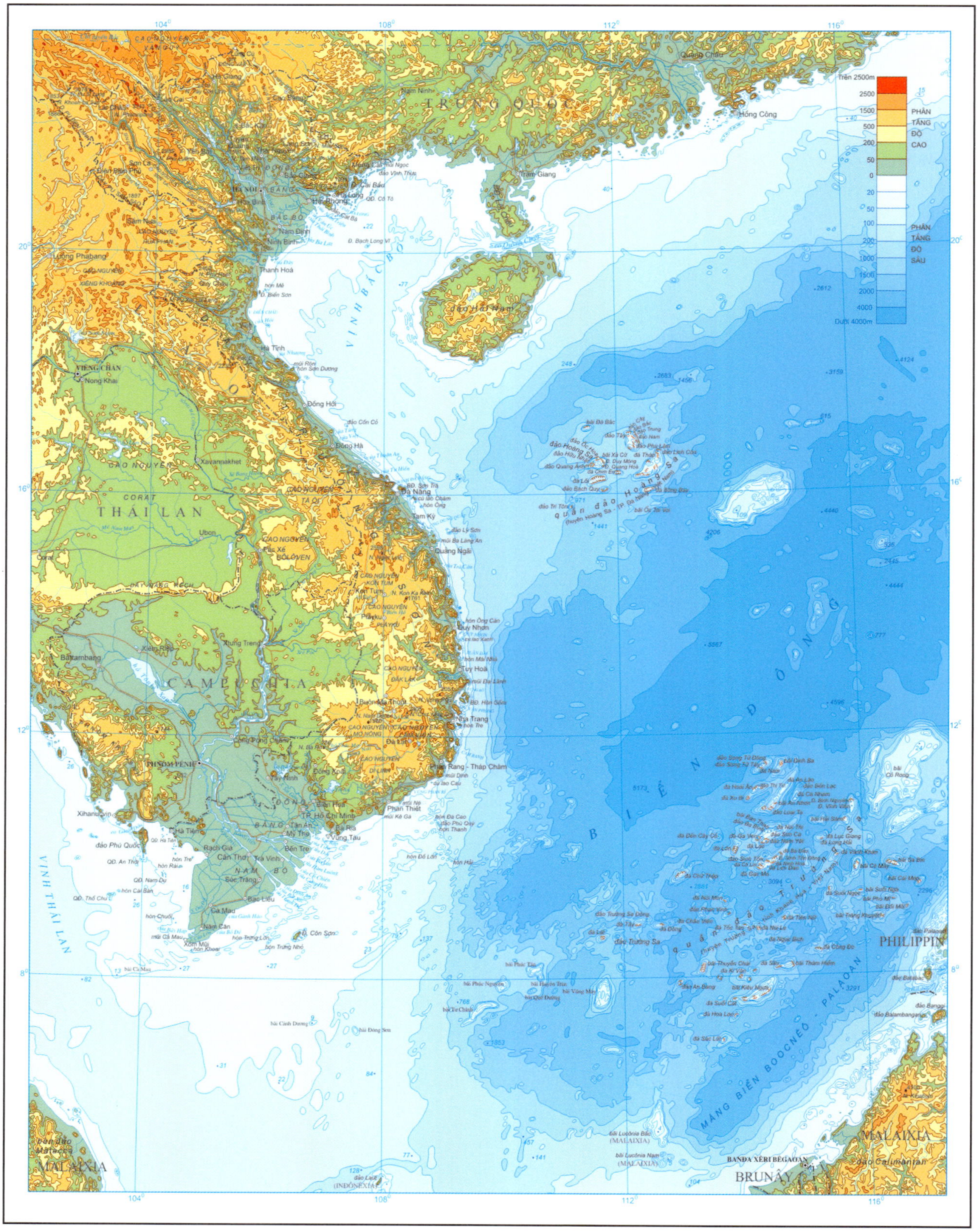

FIGURE 1. Topographic map of Vietnam and and surrounding areas of south-east Asia. Reproduced by permission from the Geographic Atlas of Vietnam (2006), Ministry of Education and Training, Education Publishing House, Hanoi.

CHAPTER 1: INTRODUCTION

Situated along the eastern part of the Indochinese Peninsula, Vietnam (the Socialist Republic of Việt Nam) is a country with a wonderfully warm coastline, a country with tropical rainforests, a country with towering mountains, a country with a turbulent past but a great future, and a country with a generous and friendly people. And Vietnam has much more to offer—including a unique cycad flora, the theme of this book.

Our first chapter consists of background information about Vietnam—its history, geography, climate, flora and fauna, ecology and vegetation types.

HISTORY

According to fossil evidence, human beings have inhabited south-east Asia for about 1.5 million years. The first traces of modern humans in Vietnam have been found in what is now Thanh Hoa Province and may date back to 30 000 years ago. Undisputed evidence established that humans were living in caves and rock shelters in the limestone hills lining the Red River Delta somewhere between 18 000 years and 9 000 years ago (Sterling et al. 2006). The early civilisations were small regional kingdoms, most of which fell under Chinese influence between 150 BC and 150 AD. Vietnam gained independence from China in AD 939, establishing its own dynasties (the Ly, 1009–1225, and the Tran, 1225–1400), then itself invading the Cham Kingdoms to the south in the late 14th Century. European merchants and missionaries had a marked influence from the 16th Century onwards, with the French Treaty of Protectorate taking effect in 1883. At that time the name of Vietnam disappeared from the world map, to reappear only after the Declaration of Independence announced by President Hồ Chí Minh on 2 September 1945. Resistance to French reoccupation, led by Hồ Chí Minh, ultimately led to the French garrison at Dien Bien Phu being overrun in 1954, with the Geneva Accord then providing for a division of the country into northern (Democratic Republic of Vietnam) and southern (Republic of Vietnam) zones at the Ben Hai (River) line. A bitter struggle then commenced between the pro-communist north and the anti-communist south, the latter gaining military support from the USA, Australia, New Zealand, Thailand and South Korea in 1965. Full-scale war ensued, with the Vietnamese Army ultimately prevailing when the Saigon forces surrendered in 1975. The subsequent reunification of the country, embracing the Đổi mới (Renovation) philosophy, allowed good progress in Vietnam's economic restructuring, with significant advances in internal stability and international co-operation.

GEOGRAPHY

Vietnam is an S-shaped country stretching approximately 1650 km from north to south but is only about 50 km wide in the central region. With a land area of 330 991 km^2 and a coastline of about 3400 km along the East Sea (South China Sea), the country is slightly larger than Italy but a little smaller than Japan. Vietnam has 59 Provinces (tỉnh, singular and plural) and 5 cities (Ha Noi, Hai Phong, Da Nang, Ho Chi Minh City and Can Tho). China is Vietnam's northern neighbour while Laos, in the centre, and Cambodia, to the south, share the western border. The population of Vietnam reached 81.4 million in 2005 and population growth continues at 2.3% annually. Hanoi (population >3 million) is the capital but Ho Chi Minh City (formerly Saigon) (population >6 million) is the largest city. Most of the inhabitants are ethnic-Viet (88%) with a small percentage being ethnic-Tay, Thai, Muong and Hoa; other minor groups are the Khmers, Chams and Montagnards. The population has a high literacy rate (about 93%) and a strong sense of religion shaped by Buddhist or Christian influences.

About 75% of Vietnam is hilly to mountainous. The country's highest peak, Fan Si Pan (Phăng Xi Păng or Hoàng Liên Sơn, 3143 m), is in the extreme northwest. The two main floodplains, used extensively for agriculture, are the Red River Delta (15 000 km^2) in the north and the Mekong Delta (39 000 km^2) in the south, the latter being one of the world's most important rice-producing zones. Soil fertility in both deltas is dependent on periodic silt deposits from alluvial flooding. The central highlands, also important agriculturally, have

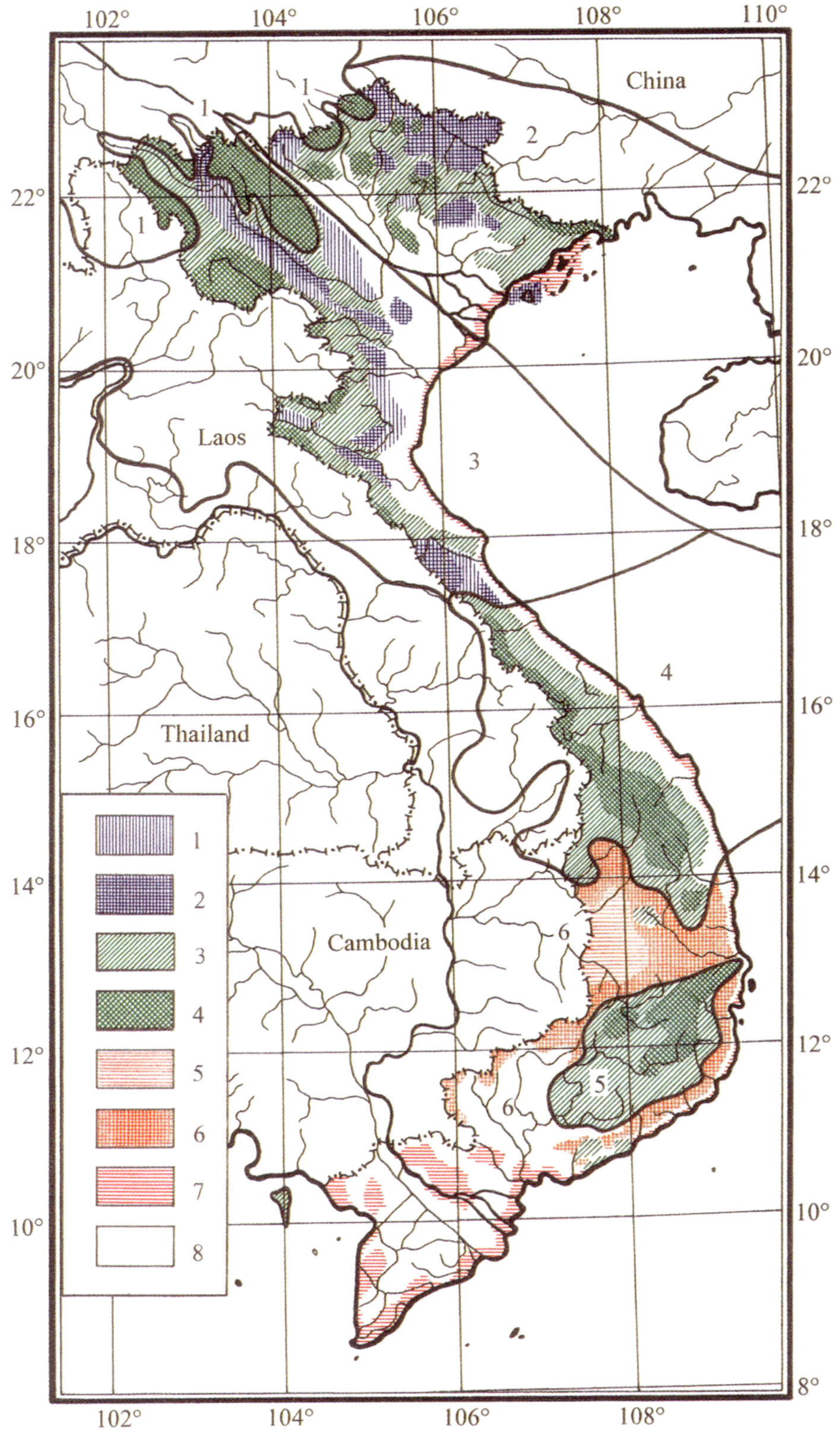

FIGURE 2. Main vegetation patterns and floristic regions of Vietnam. *Vegetation types*: 1 - Evergreen broad-leaved plain forests on alkaline soils; 2 - Evergreen and semideciduous broad-leaved, mixed and coniferous limestone mountain forests; 3 - Evergreen lowland forests on silicate rocks from sea level to 1000 m; 4 - Evergreen montane and highland forests on silicate rocks at 1000-3000 m; 5 - Semideciduous dry lowland forests; 6 - Deciduous dry lowland forests and savanna-like woodlands; 7 - Coastal vegetation, lowland wetlands and mangrove thickets; 8 - Secondary, weed and agricultural plant communities, forestry and industrial plantations. *Floristic regions*: 1 - Sikang-Yunnan Province; 2 - South Chinese Province; 3 - North Indochinese Province; 4 - Central Annamese Province; 5 - South Annamese Province; 6 - South Indochinese Province. From Averyanov et al. 2003.

rich, red volcanic soil, especially in the western parts.

The coastal ranges near Nha Trang and Da Nang (e.g. the Hai Van Pass area) are of granitic origin. In the northern parts, especially in Ha Long Bay and near the Tam Coc Caves, one encounters numerous limestone mountains, these karst formations typically, and dramatically, rising vertically from the surrounding level plains or seabed.

CLIMATE

Arising from its north-to-south range of latitudes and its topography, Vietnam's climates range from tropical with cold winters in the far north to subequatorial in the extreme south. The climate is influenced largely by the south-western monsoon winds which blow across the Gulf of Tonkin bringing warm humid weather, with occasional typhoons, to most of the country from April/June to October. Annual rainfall ranges from 1200 to 3000 mm. The period from November to February is relatively cool and dry, with temperatures falling to 5°C in the north, while the months from March to May are typically hot and dry, reaching highs of about 37°C.

FLORA AND FAUNA

Vietnam is considered as one of the most biologically diverse countries in the world and is widely recognized that the country has a globally significant proportion of rare and endemic species of plants and animals. The country has over 270 species of mammals; the forests are the home to many large animals like bears, deer, elephants, leopards and tigers, while smaller animals such as hares, monkeys, otters and squirrels occur in large numbers throughout the country. The country's reptile fauna, about 180 species, includes crocodiles, lizards, snakes, tortoises and turtles, while there are approximately 80 species of amphibians. Birds are prolific with over 800 species recorded. There are hundreds of species of fish and countless invertebrates.

With the increased access to previously unexplored areas, scientists have located many novel animal and plant species, including several that are endangered. In the 1990s, for example, the unusual bovine mammals called sao la (the Vu Quang ox, *Pseudoryx nghetinhensis*), mang lớn (the Gi Muntjac, *Megamuntiacus vuguangensis*) and mang trường sơn (the Truong Son Muntjac, *Caninmuntiacus truongsonensis*), were all discovered in the Truong Son (Annam) Mountains along the border with Laos.

The flora of Vietnam has an estimated 12 000 native higher vascular plant species, of which about 10 000 in more than 2260 genera and 307 families (including 750 introduced and cultivated species) have been identified and catalogued (Phan Ke 1998). The ten largest families are the Orchidaceae, Fabaceae, Poaceae, Euphorbiaceae, Cyperaceae, Asteraceae, Lauraceae, Fagaceae and Acanthaceae. The proportion of endemics has been variously reported, but the number of about 10–12% is largely accepted as the best estimate. The highest levels of endemism are observed in the families Acanthaceae, Anacardiaceae, Annonaceae, Apocynaceae, Araceae, Celastraceae, Ericaceae, Euphorbiaceae, Fagaceae, Myrsinaceae, Orchidaceae, Arecaceae, Rubiaceae, Sapindaceae, Theaceae and Zingiberaceae (Regalado et al. 2003a,b). A total of 356 plant species, mostly higher vascular plants, was selected as the best-known threatened taxa for listing in the Red Data Book of Vietnam (1996).

The increased international exposure of Vietnam's flora and fauna has brought with it an awareness of conservation issues. Vietnam became a signatory to the Convention in Trade in Endangered Species of Flora and Fauna (CITES) on 20 April 1994. The management authority appointed for the implementation of CITES policies is the Forestry Protection Department, Ministry of Agriculture and Rural Development; the two scientific authorities are the Institute of Ecology and Biological Resources of the Vietnamese Academy of Science and Technology, and the Centre of Resources and Environment of the Vietnam National University in Hanoi.

Vietnam has now made significant progress in nature and cultural conservation initiatives: the country now has a network of 126 protected areas including 27 national parks, 49 nature reserves, 11 species/habitat reserves and 39 land/seascape areas. National Parks and Nature Reserves include the Ba Be, Cat Ba, Ha Long Bay and Cuc Phuong National Parks in the north; Ben En, Phong Nha-Ke Bang and Bach Ma National Parks near Huế;

Nui Chua and Yok Don National Parks in the country's centre, Cat Tien, Phu Quoc, Con Dao National Parks, and Binh Chau-Phuoc Buu Nature Reserve, in the south. The total area conserved is about 25 400 km^2 and equates to approximately 7.7% of the country's land surface (Vietnamese Government publication 2003).

In addition, the Vietnamese government is addressing environmental issues through national reforestation projects (close to 1 million ha. of forest has been replanted in the past few years) and environmental educational programmes. Further impetus to a commitment to conservation is the recognition of 15 plant species or groups of species with exploitation and use strictly prohibited, and 37 other species or groups of species assigned limited exploitation and use (Vietnamese Government Decree 2006). In the same Decree, 62 threatened, rare and precious animal species have exploitation and use strictly prohibited while 89 species have been assigned to only limited exploitation and use. Ex-situ conservation of animals via captive breeding programmes is being considered and attempts are underway to maintain the medicinal plant harvest at a sustainable level.

Although positive steps have been taken towards improving conservation of flora and fauna in Vietnam, there remains a vigorous export of contraband specimens, mainly to China, through hundreds of small boundary gates. This is proving very difficult to control effectively. Trade in wild animal species involves tortoises, freshwater turtles, snakes, pangolins, civets, langurs, gibbons and bears, while the illegal plant trade includes orchids, cycads, agarwood (the fragrant and valuable heartwood from the genus *Aquilaria*) and a host of plant parts used in medicinal preparations. Clearly, all countries in south-east Asia, not just Vietnam, need strictly to enforce trade bands and to police poaching.

ECOLOGY AND VEGETATION TYPES

The diversity of Vietnam's present fauna and flora is remarkable, given the large areas of country subjected to localised physical damage, extensive logging practices and habitat destruction following the widespread use of Agent Orange (a mix of 2,4-dichlorophenoxyacetic acid and 2,4,5-trichlorophenoxyacetic acid in a solvent matrix) and similar herbicides during the north-south conflict in the 1960s and 1970s. Alarmingly, some 65 000 tons of herbicides were used in Vietnam in the period 1962–1971.

Added to those unfortunate, but relatively short-term problems, are the environmental difficulties posed by a burgeoning population with ever-increasing demands for housing, industry, mining and agriculture. Other national issues include the consequences of deforestation, slash-and-burn agriculture, logging, the medicinal plant harvest, uncontrolled hunting and local consumption of wild animals. Air and water pollution problems are severe in the larger cities. Of these various concerns, habitat loss through deforestation is identified as the most serious.

Because of environmental degradation, much of Vietnam's present vegetation exists as a complex mosaic of secondary communities. The flora contained mainly endemic elements, mixed with those from China, India and Myanmar, Indonesia and Malaysia. Mountainous areas are typically more wooded with both evergreen and deciduous forests. Woodlands predominate on the lower slopes with shrubs, herbaceous plants and grasslands in the flatter areas, all of which are considered as secondary communities. The coastal sandy zones hold complex floristic mixes, while Vietnam's southern tip, the Ca Mau Peninsula (Mui Bai Bung), is covered by dense jungle and mangrove swamps.

CHAPTER 2: CYCAD BIOLOGY

CYCAD DISTRIBUTION AND RELATIONSHIPS

Cycads, members of the plant order Cycadales, date back a remarkable 280 million years and hold a vital position in the jigsaw puzzle of plant evolution. The late Knut Norstog, pre-eminent cycad researcher, used the analogy that cycads are the 'Rosetta Stone' of seed plant evolution (Norstog & Nichols 1997). These plants have many unique features (see later in this Chapter), some of which are ancestral (primitive) and some that are derived (advanced). The order is relatively small, with only about 300 species distributed amongst 11 genera in three, possibly four, families: the Cycadaceae, the Stangeriaceae/ Boweniaceae and the Zamiaceae. Since cycads produce naked ovules (in contrast to the 300 000-odd species of flowering plants), they are technically classified with other, but unrelated, gymnosperms such as the conifers, *Ginkgo*, *Welwitschia* and several extinct groups.

Because of their ancient origin, there is a popular view that cycads are 'living fossils'–some plants are even sold under the label 'dinosaur food'. This creates the incorrect impression that cycads have remained static since their origin. In truth, the extant cycad genera have survived over many millions of years through an ability to adapt to changing conditions and a capability to find environmental niches in which they can compete successfully with other plants. By contrast, many extant cycad species may be of fairly recent origin.

Although cycad fossils have been found in all continents, including Antarctica, the present-day cycad flora is confined to the world's tropical, subtropical and near-subtropical regions. Key centres of diversity are (1) central and southern America (genera *Ceratozamia*, *Chigua*, *Dioon*, *Microcycas* and *Zamia* – all within the Zamiaceae), (2) southern and central Africa (genera *Encephalartos* in the Zamiaceae and *Stangeria* in the Stangeriaceae), and (3) Asia and Australia (genera *Cycas* in the Cycadaceae, *Bowenia* in the Stangeriaceae/ Boweniaceae, and *Macrozamia* and *Lepidozamia* – both in the Zamiaceae). Figure 3 is a cladogram summarising this information and showing the probable relationships amongst the groups.

FAMILY CYCADACEAE AND GENUS *CYCAS*

With about 100 species, *Cycas* L., the sole representative of the Cycadaceae, is the oldest, largest, most widespread, and hence the most botanically important cycad genus. Mainly arborescent plants with large compound leaves and cylindrical trunks covered with an armour of leaf bases, *Cycas* species are superficially palm-like but they differ greatly from palms in their structure and reproductive processes. The genus name was coined in 1753 by Linnaeus from the Greek word χμχας (*kukas* or *koïkas*), which had apparently been used by the Greek philosopher Theophrastus for the Egyptian doum palm *Hyphaene thebaica* (L.) Martius; the name becoming corrupted to *kykas* and later *cycas*. The type for the genus is the Indian species *C. circinalis* L.

The Cycadaceae is separated from the other cycad families on the basis of shared characters (synapomorphies) as shown in Table 1.

These features, in combination, distinguish the Cycadaceae, and hence the genus *Cycas*, from the Stangeriaceae/Boweniaceae and the Zamiaceae.

TABLE 1. Characters shared by the family Cycadaceae

- **The female reproductive structure is a loose aggregation of sporophylls**
- **Leaflets have a distinct midrib**
- **Seeds are bilaterally symmetrical**
- **Ovules are directed outwards**
- **Stipules are absent**
- **Leaflets are circinate on emergence**

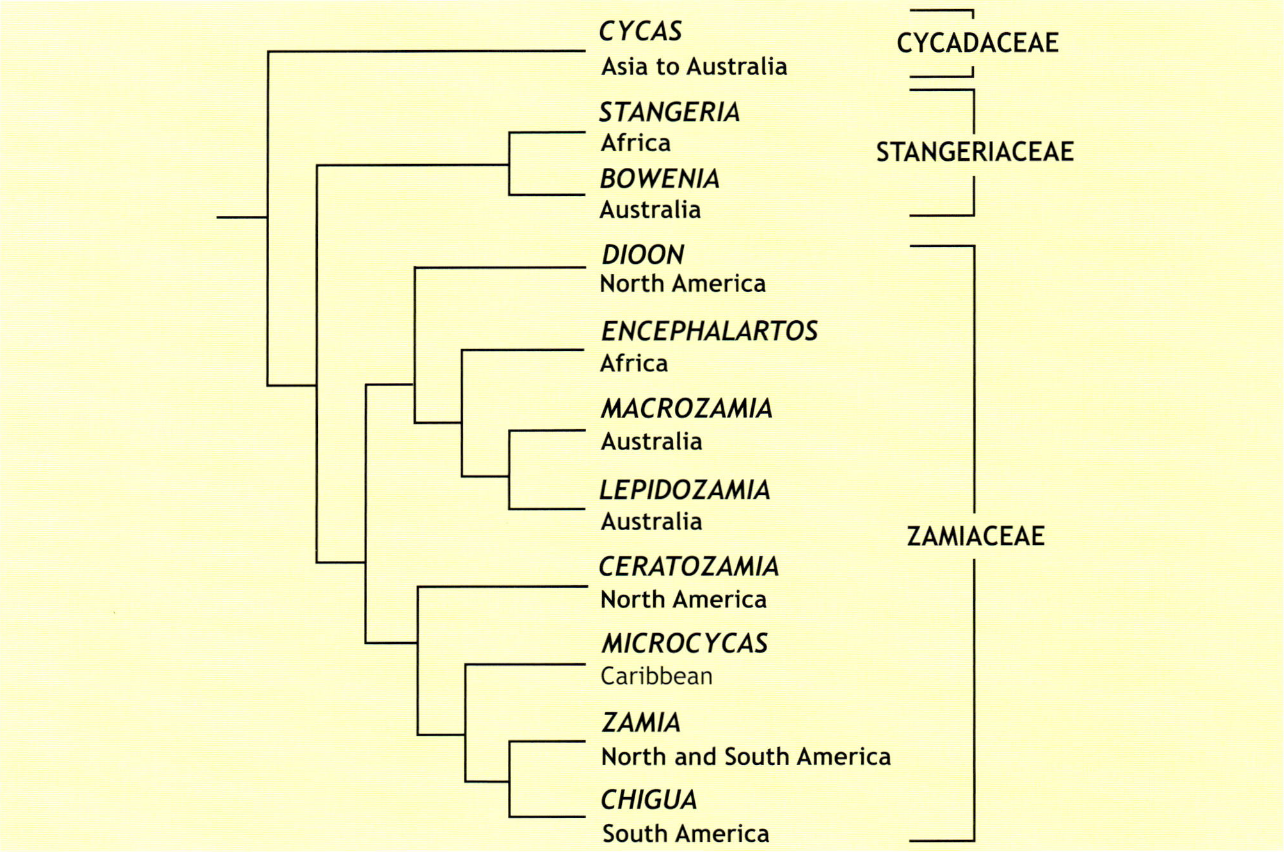

FIGURE 3. Cladogram showing probable relationships in the extant cycad genera and families. Reproduced from Hill & Osborne (2001) by permission.

ORIGIN AND SPREAD OF *CYCAS*

The question as to the origin of the first cycads poses many difficulties. Plant fossils are usually incomplete and it is rare to be certain of their position in the complex picture of evolutionary relationships. However, there is considerable support for the view held by Mamay (1976), and other palaeobotanists, that the ancestral cycads arose from the now-extinct seed ferns in the Carboniferous Period, some 300 million years ago. One of the best clues may be found in a fossilised megasporophyll named *Crossozamia* Pomel (previously also known as *Primocycas*), remarkably similar to the same structure in some present-day *Cycas* species, and found in Lower Permian rock strata near Taiyuan in north China (Gao & Thomas 1989). But there is debate as to whether this ancestral cycad originated in that area, or possibly rafted across the ancient Tethys Sea from southern Pangaea on slow-moving landmasses, or migrated across a series of landbridges as sea levels changed (Tang 2004).

What is clear is that, after the splitting of the Pangaean supercontinent into the great northern (Laurasia) and southern (Gondwana) landmasses, cycads proliferated throughout the Mesozoic Era (245–66.4 million years ago), increasing in diversity and range in response to changes in topography and climate (Osborne 2002). Plants akin to *Cycas*, however, seem to be restricted to Laurasia.

The Cenozoic Era (66.4 million years ago to the present) was a time of great change in the geography of south-east Asia. India rafted northwards to collide with southern Asia, this causing massive geological upliftment and a huge eastward movement of the Indochina block. This resulted in closure of a large gulf in the area of the present Red River, between Indochina and China. Tang (2004) speculates that this area, the 'Red River Gulf', then became a biological refugium during the freezing temperatures of the Oligocene. Further movement of landmasses along the Red River fault line would allow genetic exchanges as plant populations came together, and new speciations as habitats became isolated. Such activity may well explain the present unusually high diversity of *Cycas* in this area.

The present day distribution of *Cycas* extends from the south of Japan, southern China, Taiwan, the mainland countries of south-east Asia (Vietnam, Laos, Cambodia, Thailand); west through Myanmar, southern India, Sri Lanka, some Indian Ocean Islands, Madagascar and the eastern coast of Africa; south through the Malay Archipelago, the Philippines, Indonesia, New Guinea, many islands in the western Pacific, and further south to the northern parts of Australia. This largely oceanic distribution indicates that many of the newer *Cycas* species were dispersed under the influence of ocean currents. This overall range is enormous, extending almost 13 000 km from east to west, and over 6 000 km from north to south.

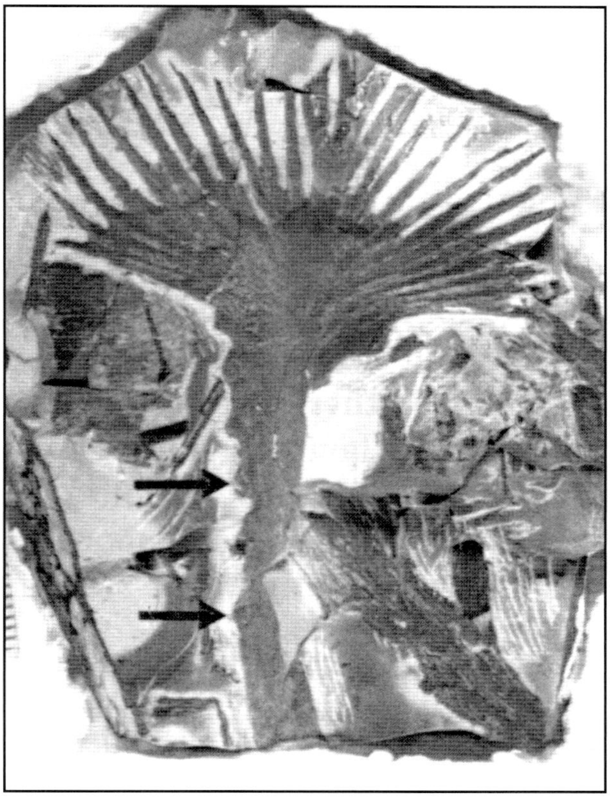

FIGURE 4. *Crossozamia* Pomel, a fossilised cycad megasporophyll from the Lower Permian in north China. Reproduced from Gao & Thomas (1989) by permission.

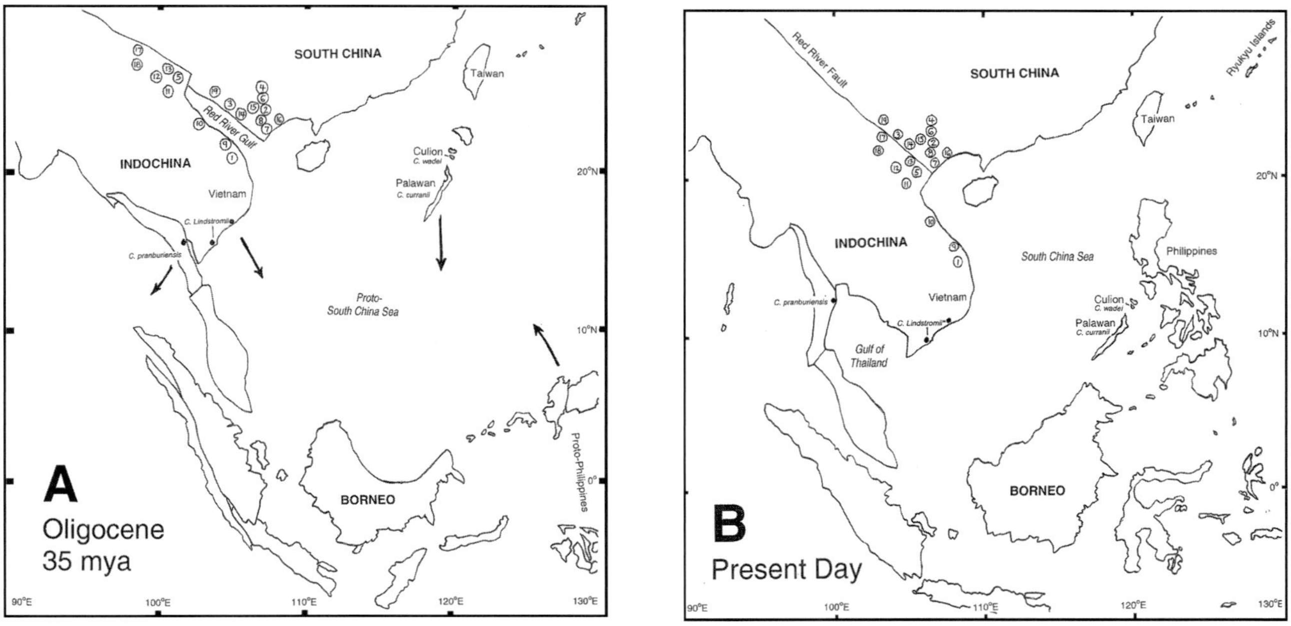

FIGURE 5. (a) Positions of the south-east Asian landmasses during the Oligocene period, 35 million years ago; arrows indicate direction of movement of continental fragments; encircled numbers indicate corresponding localities for present-day cycad species. (b) Present-day landmasses in south-east Asia; encircled numbers indicate location of the corresponding species in (a), showing how continental drift may have moved them. Maps reproduced from Tang 2004, after Hall 1998, by permission.

CLASSIFICATION OF *CYCAS*

Various authors have proposed systems to group the many species of *Cycas* in some organised fashion. The most noteworthy of such infrageneric classifications are those developed in a series of publications by Ken Hill (e.g. Hill et al. 2004), in which all available morphological and molecular evidence is pooled and statistically data-processed to find the most probable set of relationships between species. Using this approach, Hill recognises 5 sections in the Cycadaceae, keyed out as indicated in Table 2. Further useful discussion on the matter of selection of morphological characters that can be used to determine species boundaries is made by Lindström (2004).

All naturally occurring Vietnamese cycads fall within three sections, as shown in Table 3.

TABLE 2. Key to the sections of the genus *Cycas* (after Hill et al. 2004)

1. Ovules tomentose	**Section *Asiorientales***
1. Ovules glabrous	**2**
2. Megasporophyll blade pectinate	**3**
2. Megasporophyll blade not pectinate	**Section *Cycas***
3. Males cones soft, microsporophyll apices not deflexed	**4**
3. Male cones rigid, waxy, microsporophyll apices deflexed	**5**
4. Sclerotesta not ribbed	*Cycas panzhihuaensis*
4. Sclerotesta ribbed	**Section *Wadeanae***
5. Microsporophylls soft and rounded, sarcotesta not fibrous, sclerotesta wrinkled	**Section *Stangerioides***
5. Microsporophylls hard and sharply-pointed, sarcotesta fibrous, sclerotesta smooth	**Section *Indosinenses***

TABLE 3. Section and subsection assignments for the Vietnamese *Cycas* species
(Numbers refer to the sequence of treatment in the descriptive section of this text)

Section *Stangerioides*	**Section *Indosinenses***	**Section *Cycas***
1. *C. collina*	19. *C. pectinata*	**Subsection *Cycas***
2. *C. fugax*	20. *C. clivicola* subsp. *lutea*	26. *C. inermis*
3. *C. simplicipinna*	21. *C. elongata*	
4. *C. balansae*	22. *C. pachypoda*	**Subsection *Rumphiae***
5. *C. aculeata*	23. *C. condaoensis*	27. *C. litoralis*
6. *C. tanqingii*	24. *C. lindstromii*	
7. *C. chevalieri*	25. *C. siamensis*	
8. *C. hoabinhensis*		
9. *C. brachycantha*		
10. *C. dolichophylla*		
11. *C. micholitzii*		
12. *C. bifida*		
13. *C. multipinnata*		
14. *C. ferruginea*		
15. *C. sexseminifera*		
16. *C. tropophylla*		

Cycas species from China that are expected to occur in Vietnam (both are in **Section *Stangerioides***):

17. *C. segmentifida*
18. *C. diannanensis*

DISCOVERY OF *CYCAS* SPECIES IN VIETNAM

Although the Vietnamese people have been aware of their cycad heritage for many generations, it is only from the 18th Century onwards that any knowledge of these plants reached the western world. The first two cycads to be described were *Cycas circinalis* L. from India in 1753 and *C. revoluta* Thunb. from Japan in 1782. In 1793, the Jesuit priest and naturalist João de Loureiro published a description of a third cycad, *Cycas inermis*, based on material he had collected in southern Vietnam (then Cochin China). Two more species, *C. tonkinensis* and *C. bellefonti* which had been collected in northern Vietnam (then Tonkin) and subsequently cultivated in European gardens, were described by Linden and Rodigas in 1885 and 1886 but no herbarium specimens were made and the names are no longer considered valid. In 1900, Otto Warburg described *C. balansae* from near Hanoi; five years later William Thiselton-Dyer published the species *C. micholitzii* from central Vietnam-Laos (then Annam). In 1931, Jacques Leandri described *C. chevalieri* from northern Vietnam and *C. pectinata* var. *elongata* (now *C. elongata*) from southern Vietnam. The widespread south-east Asian species *C. pectinata* Buch.-Ham. and *C. siamensis* Miq. were based on material from India and Thailand, although these species are both known from Vietnam.

The names *Cycas circinalis* L., *C. rumphii* Miq., *C. undulata* Desf. ex Gaudich., *C. miquelii* Warb. and *C. immersa* Craib have been incorrectly applied to Vietnamese cycads. A proposed segregation of some *Cycas* species into a new genus, *Epicycas*, as proposed by de Laubenfels and Adema in 1998, has not met with approval from cycad taxonomists (Chen et al. 2004) and is not supported by molecular studies (Kokubugata & Forster 2006).

For most of the 1900's, the total number of Vietnamese cycad taxa was thought to be 10 or less; the 1996 *Flora du Cambodge du Laos et du Viêtnam*, for example, listed only 8 species (Hiêp & Vidal 1996). However, during the last 12 years extensive fieldwork has been done on the Vietnamese cycad flora. The efforts of local botanists Hiêp Nguyen and Loc Phan Ke, Chinese scientists Si Lin Yang, Ding Yue Wang, Fan Nan Wei and Chia Jui Chen, the Thailand resident horticulturist Anders Lindström, and most particularly the Australian taxonomist, Ken D. Hill, have now led to the publication of an astonishing 22 species over the period 1994–2004, bringing the number of Vietnamese cycads to the present total of 27 (Hiêp & Loc 1997, 1999; Hill et al. 2004). An inspection of the literature reveals that 39 of Vietnam's 58 Provinces have populations of one or more native cycads. The distribution of these species, in terms of their sections, is shown in Fig. 7.

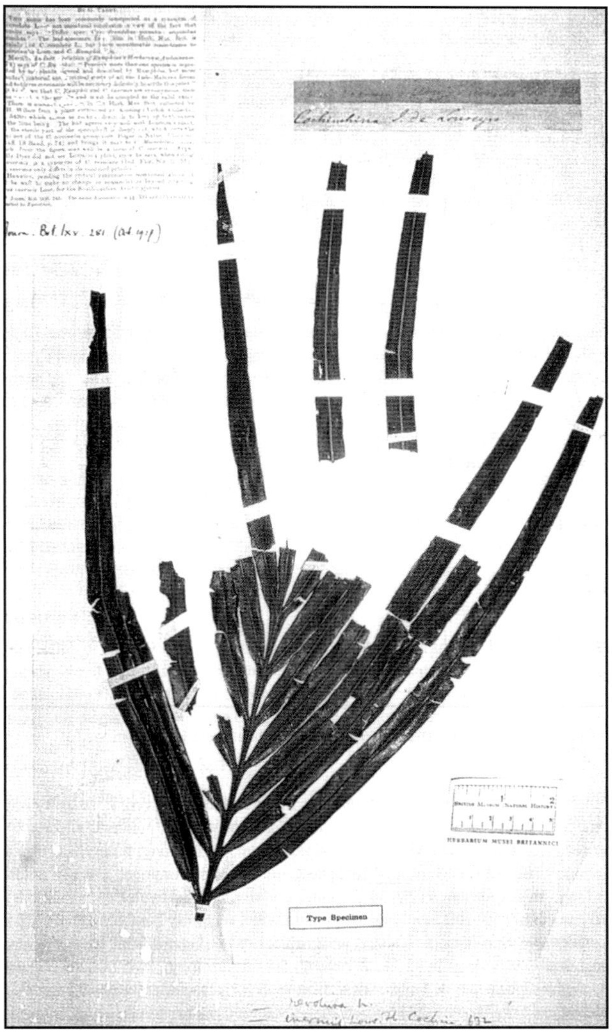

FIGURE 6. Holotype of *Cycas inermis* collected by João de Loureiro for the species description in 1793, as filed in the herbarium of the Natural History Museum, London. Reproduced from Hill et al. (2004) by permission.

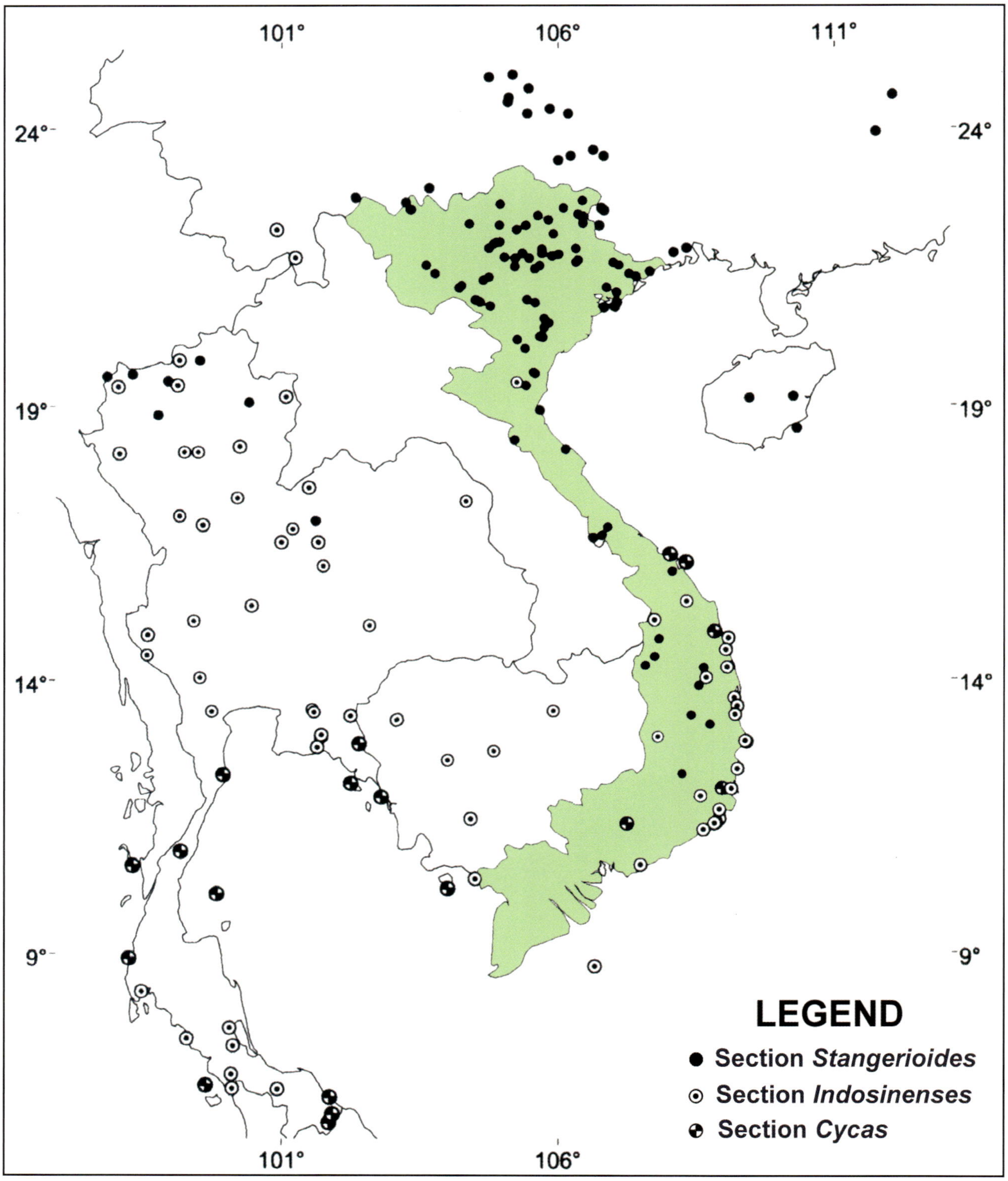

FIGURE 7. Map of *Cycas* sections in Vietnam and surrounding areas in south-east Asia. Reproduced from Hill et al. (2004) by permission.

VEGETATIVE MORPHOLOGY

All *Cycas* species have a stout trunk; in some species (e.g. *C. siamensis*) this can reach 12 m in height but in others the trunk is almost wholly subterranean (e.g. *C. lindstromii*, *C. micholitzii*, *C. simplicipinna*). Anatomically, the trunks are termed pachycaulous, i.e. they comprise mainly soft starch-rich storage tissue with very little true wood. Veins, leading to the leaves, run longitudinally up the stem, positioned in a circle as seen in cross section; the girdling leaf traces are known also from some ferns but no other seed plants. Cycad trunks are sometimes greatly swollen in an 'elephant's foot' fashion towards ground level (e.g. *C. pachypoda*, *C. condaoensis*). In older specimens of some species, trunks may be branched (e.g. *C. elongata*); such branching is adventitious, as opposed to axillary, and may be dichotomous. Adventitious branching at ground level leads to 'suckers' which eventually give a multi-trunked appearance to a single plant. Trunk surfaces are initially clothed in an armour of old leaf bases and cataphyll remnants but these become lost with age; in old plants the bark becomes deeply fissured and presents a fascinating texture often in a tortoise-shell pattern.

Leaves of *Cycas* are always compound, either once pinnately divided (most species), twice (or more) pinnately divided (e.g. *C. multipinnata*), or with dichotomously divided leaflets (e.g. *C. bifida*, *C. micholitzii*). Leaves are generally a mid- to deep green glossy upper surface and a somewhat paler and less glossy lower surface (i.e. they are discolorous). Leaves of most species are in the range 150–500 cm long but some species bear leaves up to an impressive 7 m in length (e.g. *C. bifida*). Leaves may be strikingly and densely tomentose on emergence (e.g. *C. ferruginea*); in some cases this tomentum persists to maturity (e.g. *C. siamensis*). Leaves may be very few in number (e.g. *C. fugax*) or may exceed 50 per crown (e.g. *C. pachypoda*). Leaves may be flat (e.g. *C. inermis*) or may be strongly keeled (e.g. *C. tropophylla*). Leaves are generally persistent from one season to the next, but plants in drier, colder, inland areas may become deciduous for short periods (e.g. *C. siamensis*).

Leaflets generally reduce in size towards each end of the leaf. Towards the base, the leaflets may reduce abruptly, or gradually, to a series of spines

FIGURE 8. *Cycas pachypoda*, trunk detail, plant from Vinh Hoa village, Binh Thuan Province. Photo: Ken Hill.

FIGURE 9. *Cycas dolichophylla*, leaf detail showing highly glossy leaflets with undulate margins, from a plant in cultivation, Be Nuoc village, Cao Bang Province. Photo: Ken Hill.

along the leaf stalk (petiole). The petiole may thus be covered with such spines for nearly all its length, some lesser proportion, or it may be unarmed. This character, however, is very variable and is not diagnostic at the species level. Petioles may be glabrous (e.g. *C. tanqingii*) or pubescent (e.g. the dark purple-brown tomentum of *C. balansae*).

Cycas leaflets have a prominent midrib, a character that the Cycadaceae share with *Stangeria* and *Chiqua*. However, the *Cycas* midrib holds a single vascular bundle, whereas that in the other two genera is a compound trace of numerous veins. *Cycas* leaflets, but not leaves, emerge in a tightly coiled (circinate) manner, a character that further separates *Cycas* from other cycad genera. Leaflets of *Cycas* are always without lateral thorns or prickles (i.e. they are entire) but may be undulate along their margins (e.g. *C. aculeata*, *C. dolichophylla*). Leaflets may be straight (e.g. *C. tropophylla*) or gently curved (e.g. *C. condaoensis*). *Cycas* leaflets are also described as being decurrent, i.e. leaflet blades extend slightly downwards towards the leaf base, along the supporting rachis.

In addition to true leaves, cycad stem apices produce bract-like structures (cataphylls), alternately with the true leaves and reproductive structures. These are generally short, soft, papery or leathery, typically about 60 mm long, often tomentose, usually non-photosynthetic structures which serve to protect the subsequently emerging softer leaves or cones.

The primary cycad roots are soft, stubby, relatively short structures that serve the usual purposes of anchoring the plant and extracting soil nutrients. A different root system, seen in all cycads, is the negatively-geotropic highly-branched root clusters

FIGURE 10. ***Cycas sexseminifera*, showing long-lasting cataphylls, from a plant in the gardens of the Airforce Museum, Hanoi. Photo: Ken Hill.**

FIGURE 11. ***Cycas elongata*, detail of stem apex showing spinescent petioles and soft cataphylls with a sparse tomentum. Specimen 95606A at Montgomery Botanical Center, Florida, USA. Photo: Jody Haynes.**

FIGURE 12. Leaf of a *Cycas dolichophylla* x *C. bifida* hybrid, with leaflets showing the characters of both parents, from a plant near Thai Nguyen City, Thai Nguyen Province. Photo: Ken Hill.

called coralloid roots that develop just below soil level and are prone to colonisation by cyanobacteria (previously called blue-green algae); this resulting in a nitrogen-fixing symbiosis like that known in legumes. This partnership may be an important survival strategy for cycads growing in a nutrient-poor environment.

REPRODUCTIVE FEATURES

All extant members of the Cycadales bear male and female reproductive structures on separate plants (i.e. they are dioecious). Spontaneous sex change is known to occur occasionally in most cycad genera, but only in cultivated plants and usually associated with some environmental distress (Osborne & Gorelick in print).

Male *Cycas* plants bear one or more terminal cylindrical cones with numerous microsporophylls spiralling out horizontally from a stout central axis. Each microsporophyll has many pollen-producing microsporangia on a fertile zone on its undersurface. Microsporophylls may be soft and rounded (section *Stangerioides*) or hard and sharply tipped (section *Indosinenses*). The cones may be soft or rigid, glabrous to pubescent, and range in colour from greenish, through cream, pale yellow to dark orange. Male cones vary in shape from narrowly ovoid to spindle-shaped (fusiform) and in size in from smallish (e.g. 10–12 cm long x 2.5–4 cm diameter for *C. fugax*) to much larger (e.g. 35–45 cm long x 13–17 cm diameter for *C. litoralis*).

Male cones of *Cycas* species produce copious amounts of pollen, typically about 500 ml in volume. Under microscopic examination the pollen grains are seen to be ellipsoidal, about 30 μm in length, scored with a deep longitudinal furrow but otherwise without significant ornamentation.

Female *Cycas* plants bear a 'pseudocone', somewhat similar in shape to a cabbage. This is an indeterminate structure of loosely-arranged, highly modified leaves (megasporophylls), each with 2, 4, 6 or 8 ovules distributed on each side of a stalk supporting a spoon-shaped blade (lamina). This ancestral character is akin to that seen in fossils of seed ferns and differs substantially from the more advanced arrangement in other cycad genera. Other cycad families have megasporophylls usually bearing only two ovules; they are thickened, reduced and more tightly aggregated into a compact determinate cone similar to that of the male plants. Megasporophylls may bear small serrations (section *Cycas*) or they may be ornamented with numerous spines, some up to 9 cm in length (e.g. *C. fugax*), in a comb-like (pectinate) fashion (sections *Indosinenses* and *Stangerioides*). Megasporophylls vary from deep green in colour through to yellow or orange, with varying degrees of pubescence.

Cycas seeds are bilaterally symmetrical (platyspermic), unlike the radially symmetrical (radiospermic) profile of seeds of other cycad genera. Seeds of species in section Asiorientales are tomentose (e.g. non-Vietnamese species *C. revoluta* Thunb. and *C. taitungensis* C. F. Shen, K. D. Hill, C. H. Tsou & C. J. Chen) but all other cycads produce glabrous seeds. These are initially green, but become yellow to orange or brown with age. Typically, *Cycas* seeds are about 40 mm long and 30 mm wide at maturity. The internal structure of the seeds provides additional clues as to species relationships. An outer fleshy layer, the sarcotesta, has a fibrous texture in section *Indosinenses*, but is smooth in texture in section *Stangerioides*. The outer fleshy layer surrounds a hardened internal shell (sclerotesta), which is smooth in section *Indosinenses* but warty (verrucose) in the *Stangerioides*. Within the sclerotesta is the megagametophyte, the haploid starch and protein-rich tissue that provides nourishment to the developing embryo, the diploid product of fertilisation. One further character is seen in subsection *Rumphiae* of section *Cycas*, where *C. litoralis* is the only Vietnamese representative; these seeds have a well-developed internal spongy layer that allows them to float in water. This is consistent with the distribution of species in subsection *Rumphiae* along coastal fringes and through islands in the Indian and Pacific Oceans.

FIGURE 13. *Cycas tropophylla*. A cluster of highly dissected and tightly appressed megasporophylls emerges at the stem apex of this plant in cultivation in Hanoi. This structure is analogous to the female cone in other cycad genera. Photo: Ken Hill.

FIGURE 14. *Cycas elongata*, male cone showing the spiral arrangement of the microsporophylls, from a plant in cultivation in Ho Chi Minh City. Photo: Roy Osborne.

Right – FIGURE 15. *Cycas segmentifida*, megasporophyll with developing seeds, from a plant in cultivation at the South China Botanical Garden, Guangdong Province, China. Photo: Ding Yue Wang.

POLLINATION AND SEED DISPERSAL

At the time of pollen shedding, the axis of male cycad cones extends significantly so that the microsporophylls became separated and intense respiration commences in a diurnal rhythm. This process, called thermogenesis, results in the cone temperature peaking at a temperature several degrees above ambient, generally in the late afternoon. During the heating cycle, volatile odours are released and insect activity becomes much more obvious. Jones (2002) records that male cones of *Cycas* section *Stangerioides* have a mild, sweet scent whereas those from section *Indosinenses* have a strong, harsh scent. All this evidence supports the view that these plants have evolved a strategy to attract specific insect pollinators (entomophily).

In the mid-1990s, William Tang and Si Lin Yang visited cycad populations in Thailand, Vietnam and China to survey putative insect pollinators. Two genera of insects were found to occur consistently and abundantly on cones of Asian cycads; these were weevils of the genus *Tychiodes* Wollaston and languriid beetles of the genus *Xenocryptus* Arrow (Tang et al. 1999). Examples of *Tychiodes* were found only in southern Vietnam but *Xenocryptus* species were found in both the north and the south. Insect pollination has been clearly documented in other cycads, especially *Encephalartos*, *Lepidozamia*, *Macrozamia* and *Zamia* (all Zamiaceae) (Norstog & Nicholls 1997).

Despite all the circumstantial evidence for entomophily, there is still some support that wind pollination may be effective in at least some *Cycas* species. Supporters of this belief point to the large quantity of pollen produced, the fact that some species bear male cones well above ground level, the surface texture of the female megasporophylls, and the fact that *Cycas* ovules point outwards; all factors consistent with wind pollination (anemophily).

Following the pollination event, pollen grains produce a tube at the megagametophyte entrance; from this tube a motile sperm cell emerges (another archaic feature that cycads share with primitive plants) and it is this cell that fertilises the ovule. The resulting embryo slowly develops in size and complexity while the seed is still attached to the plant; this process takes many months—sometimes as long as two years. During this time, the megasporophylls slowly extend, curve down, wither and die, when the seeds eventually fall from the mother plant.

Little is known as to how seeds of the Asian cycads are dispersed. Small mammals, rodents and birds probably play a role. Some seeds will simply roll down slopes under the influence of gravity. Seeds of coastal species with the internal spongy layer (e.g. *Cycas litoralis*) may reach the intertidal zone and be transported by ocean currents.

FIGURE 16. Pollinators of Vietnamese cycads. (a & b) Dorsal and lateral views of *Tychiodes* Wollaston from southern Vietnam; (c & d) Dorsal and lateral views of *Xenocryptus* Arrow found in both northern and southern Vietnam. Bar = 1 mm. Photos: William Tang.

HYBRIDISM

Geographic separation is the most effective reproductive barrier between different cycad species. But when different species grow within pollination range of each other, some breakdown of reproductive isolation occurs and hybrids appear in one or both the populations. This is especially true for plants like cycads, with an apparent lack of pollinator specificity and with generally weak inherent fertility barriers. Obvious hybrids are recognised as being morphologically intermediate between the parent species and also by showing a high degree of variability, as expected from Mendelian segregation, in the second and later generations. More subtle is the slow genetic exchange between different taxa which may have occurred over many generations and which have led to intergrading between populations or species.

Examples of hybridism in Vietnam are shown in Table 5.

TABLE 5. Known cycad hybrids in Vietnam

Hybrid *Cycas* parents	Location
C. dolichophylla x *C. ferruginea*	Eastern Thai Nguyen Province
C. collina x *C. dolichophylla*	West of Thai Nguyen, Son La Province
C. dolichophylla x *C. bifida*	Western Thai Nguyen Province

TOXICITY

All cycads must be regarded as potentially toxic. Chemically, the toxins are of three kinds: (a) the methylazoxymethanol (MAM) glycosides such as cycasin, macrozamin and neocycasins, (b) amino acids such as β-methylamino-L-alanine (BMAA) and β-N-oxalylamino-L-alanine (BOAA), and (c) unusual steryl glucosides. The toxins appear to be concentrated in the seed, particularly in the megametophytic tissue, and various accounts have been put forward in which it is proposed that these compounds have caused acute and/or chronic poisoning to humans or livestock. It is well known that various peoples such as the Australian Aboriginals, several African tribes, some central American folk and some Asian groups have used cycad seed as a food supplement, most often in times of famine. Significantly, in each case, the seed is detoxified by some kind of washing or roasting process before consumption.

Medical interest has focussed particularly on the high incidence of neurological conditions such as amyotrophic lateral sclerosis (ALS) Parkinson's disease (PD) which manifested after World War II on the western Pacific island of Guam. Epidemiological data support a link between the disease incidence and the consumption of flour made from the Guam endemic cycad, *Cycas micronesica* K. D. Hill. These studies continue to generate sometimes-controversial hypotheses as to the nature of the responsible agent and its mode of action (e.g. Sacks 1996, Cox in print; Shaw et al. in print).

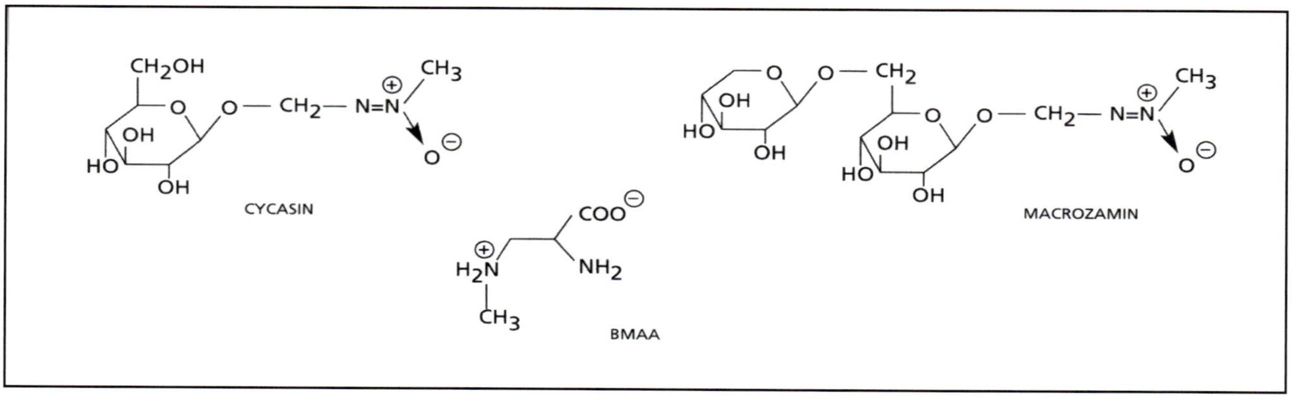

FIGURE 17. Chemical structures of cycad toxins cycasin, macrozamin and BMAA. Reproduced from Hill & Osborne (2001) by permission.

ETHNOBOTANY

Vietnamese people hold cycads in high esteem as ornamentals. Specimens of the Japanese *Cycas revoluta* Thunb. have been used extensively in pagodas, temples, public and private gardens, either as in-ground plantings or as prized container plants. It is thought that these plants originated in China and it is curious that almost all plants are female (Hiêp & Loc 1999). Since the 1980s, a new trend has recently emerged where native Vietnamese cycads are being used in pagodas, public gardens and in private homes as feature specimens; those with unusual stem and foliage features have become increasing popular (Hiêp & Loc 1999). This has led to a much greater demand for wild-collected plants and has resulted in increased pressure on native populations (see the Conservation section, below).

The Vietnamese name for the genus is tuế and *C. revoluta* is specifically known as vạn tuế (ten thousand year old cycad). Native cycads, by contrast, are called thiên tuế (one thousand year old cycad). Local names for the individual species extend these words to incorporate morphological and geographical data. For example, *C. pachypoda* is thiên tuế chân voi-chân voi (elephant leg cycad) while *C. tropophylla* is known under the local name thiên tuế Hạ Long (Ha Long cycad). Whitelock (2002) mentions an ethnic name cáy ngen for *C. chevalieri* but we are unable to verify this.

In past times, people in south-east Asia have undoubtedly used cycads for food, medicine, and possibly in religious ceremonies, but details of these usages in Vietnam are not known to the authors and not recorded in the literature.

CONSERVATION

The growing popularity of Vietnamese cycads for both local use and for export (mainly from north Vietnam to China) has resulted in many cycad populations being exploited by commercial collectors. This, coupled with the additional pressure of habitat loss, has contributed to a sometimes steep reduction in the size of many cycad populations. Decline in the number of pollinators and seed dispersers, reduced plant recruitment and the illegal collection of mature specimens, are problems which need to be addressed by conservation authorities.

The International Union for Conservation of Nature and Natural Resources (IUCN), as advised by the Threatened Plants Committee, has attempted to list all plant and animal species into categories of vulnerability in their 'Red Data Book' series. The categories include extinct (EX), extinct in the wild (EW), critically endangered (CR), endangered (EN), vulnerable (VU), near threatened (NT) and various lower risk rankings. Although only one Vietnamese endemic cycad, *Cycas micholitzii*, was listed in the IUCN 1997 Red List, all species have since been allocated provisional conservation status codes (Hill et al. 2003). At present one species is listed as critical (*C. fugax*), three are considered endangered (*C. aculeata*, *C. hoabinhensis* and *C. multipinnata*), while the others are either vulnerable or near threatened. These details are summarised in Table 6.

Although there are no sites set aside in Vietnam specifically for their conservation, cycads are present

Figure 18. Wild-collected *Cycas ferruginea* plants being offered for sale on the highway south-west of Lang Son City, Lang Son Province. Photo: Roy Osborne.

in at least 23 protected areas, as shown in Table 7.

However, officials in charge of these National Parks and Nature reserves are often unaware of the threatened status and conservation risks to cycads. Rangers are generally not well paid and can sometimes be bribed. At the present time, there are only limited and relatively ineffective efforts to protect cycads from exploitation and to prohibit export of wild-collected plants. Vietnamese conservation authorities would do well to consider establishing community-based nursery projects similar to that commenced in southern China in 1999 for the conservation of *Cycas debaoensis* Y. C. Zhong & C. J. Chen (Tang et al. 2004).

TABLE 6. Provisional conservation status of Vietnamese cycads (after Hill et al. 2003)

Cycas **species**	**IUCN status (2003)**	**Population size**	**Present decline**	**Range (km)**	**% Habitat reduction**
C. aculeata	EN	250–2500	low	20	30–50
C. balansae	NT	>10 000	low	400	<30
C. bifida	VU	>10 000	high	100	30–50
C. brachycantha	NT	>10 000	low	50	unknown
C. chevalieri	NT	>10 000	low	100	<30
C. clivicola	NT	>10 000	low	1000	<20
C. collina	VU	2500–10 000	low	200	30-50
C. condaoensis	VU	2500–10 000	stable	20	<20
C. dolichophylla	VU	>10 000	low	500	30–50
C. elongata	VU	>10 000	high	100	30–50
C. ferruginea	NT	>100 000	low	150	20
C. fugax	CR	250–2000	high	200	80
C. hoabinhensis	EN	2500–10 000	high	50	50–80
C. inermis	VU	>10 000	high	200	30–50
C. lindstromii	VU	>10 000	high	200	30–50
C. pectinata	VU	>200 000	low	3000	30–50
C. sexseminifera	NT	>10 000	low	500	<20
C. siamensis	VU	>10 000	low	1000	30–50
C. tanqingii	NT	>100 000	stable	100	<20
C. tropophylla	NT	>10 000	low	50	<20

Explanation of symbols: CR = critically endangered; EN = endangered; VU = vulnerable; NT = near threatened.

TABLE 7. Protected areas in Vietnam with cycads naturally present

Cycas species present	Name of protected area	Province/Municipality in Vietnam
C. brachycantha	Ba Be National Park	Bac Kan
C. condaoensis	Con Dao National Park	Ba Ria-Vung Tau
C. dolichophylla	Ben En National Park	Thanh Hoa
C. dolichophylla	Cuc Phuong National Park	Ninh Binh
C. dolichophylla	Na Hang Nature Reserve	Tuyen Quang
C. ferruginea	Huu Lien Nature Reserve	Lang Son
C. hoabinhensis	Chua Huong Tich Nature Reserve	Ha Tay
C. hoabinhensis	Cuc Phuong National Park	Ninh Binh
C. hoabinhensis	Thuong Tien Nature Reserve	Hoa Binh
C. inermis	Nam Cat Tien National Park	Dong Nai
C. inermis	Son Tra Nature Reserve	Da Nang
C. lindstromii	Binh Chau-Phuoc Buu Nature	Ba Ria-Vung Tau
C. litoralis	Phu Quoc National Park	Kien Giang
C. micholitzii	Deo Ngoan Muc Nature Reserve	Lam Dong
C. micholitzii	Ho Lac Nature Reserve	Dak Lak
C. micholitzii	Mom Ray Nature Reserve	Kon Tum
C. pectinata	Bu Gia Map Nature Reserve	Song Be
C. pectinata	Kon Cha Rang Nature Reserve	Gia Lai
C. sexseminifera	Ben En National Park	Thanh Hoa
C. siamensis	Nam Ca Nature Reserve	Dak Lac
C. siamensis	Yok Don Nature Reserve	Dak Lac
C. tropophylla	Cat Ba National Park	Hai Phong
C. tropophylla	Ha Long Bay Nature Reserve	Quang Ninh

FIGURE 19. Wild-collected *Cycas elongata* at Dong Dang town, Lang Son Province, apparently waiting for transportation across the nearby border to China. Photo: Roy Osborne.

FIGURE 20. Cycad habitat loss. Part of a *Cycas elongata* locality near Cam Lap village, Khanh Hoa Province, after being cleared for a banana plantation. Photo: Roy Osborne.

CULTIVATION

As mentioned under the Ethnobotany section, exotic cycads have been used in Vietnam for many years and are commonly sold in Vietnamese nurseries. However, there is now a small but increasing demand for native cycads; these are being used either as garden subjects or a feature plants in containers.

As mentioned in the Conservation section, native cycads are, unfortunately, usually removed from the wild and sold without proper authority. Buying these illegal plants encourages more such exploitation and places plant populations at greater risk and should be discouraged. In due course, the horticultural demand may lead to more acceptable cycad cultivation by nurserymen, possibly with sustained harvesting of wild seed under a controlled permit system or possibly after the establishment of community-based nurseries as mentioned above.

Cycads are very forgiving plants as garden subjects. Their needs are like most garden or container plants: good light, well-drained and good-quality soil, watering when necessary and a fertiliser application at the start of the growing season. Cycads are commonly propagated from seed, although they are fairly slow in the early growth stages. Basal suckers can be used for propagation if they are carefully separated from mother plants and steps are taken to prevent rotting. Pests are generally not a serious problem although the increasing incidence of infestations of the Cycad *Aulacaspis* Scale (CAS) is cause for concern*.

***Cycads in south-east Asia are host plants to a small, white, armoured scale insect called *Aulacaspis yasumatsui* Takagi (Cycad *Aulacaspis* Scale, CAS, Hemiptera: Diaspididae) (Haynes 2005). In wild populations, there is a balance between this insect and its natural predators. However, the insect can cause serious damage to plants in cultivation where natural predators do not keep the pest in check. CAS has recently decimated cultivated cycads in Indonesia, Florida, China, Guam and Taiwan, and it now appears to be a major threat to wild *Cycas* populations anywhere outside south-east Asia. Growers have been advised to be vigilant and take all possible steps to prevent the spread of this insect.**

Readers interested in finding out more details on cycad cultivation should consult the books by Grobbelaar (2002), Jones (2002) and Whitelock (2002), all of which have useful information on the subject.

***Cycas tropophylla* in cultivation, Cam Pha Town, Quang Ninh Province.** **See Fig 36 (g) on p. 67.**

CHAPTER 3: A KEY TO *CYCAS* SPECIES IN VIETNAM

The assignment of any particular cycad to its species is not always easy. Helpful information, not all of which is usually available, includes a knowledge of the locality of naturally-occurring populations, detailed data of the vegetative morphology and male and female reproductive structures, and sometimes anatomical and even biochemical data. A useful key to the Vietnamese cycads has been drawn up by Hill (2004) and is shown in Table 8, in slightly modified form. Although it does not occur naturally in Vietnam, the Japanese species *C. revoluta* is included in the key since it is widely cultivated as an ornamental. The two Chinese species, *C. segmentifida* and *C. diannanensis*, expected to occur in Vietnam, are not included in the key.

Table 8: A key to *Cycas* species in Vietnam (after Hill et al. 2004)

1. Leaflets not divided	
2. Megasporophylls pectinate	
3. Leaflets with margins revolute or strongly recurved	*C. revoluta*
3. Leaflets with margins flat or moderately recurved	
4. Male cone soft, sarcotesta not fibrous, sclerotesta wrinkled	
5. Megasporophyll apical spine enlarged	
6. Leaves flat	
7. Megasporophyll blade 35–55 x 30–50 mm	*C. sexseminifera*
7. Megasporophyll blade 65–120 x 60–80 mm	*C. ferruginea*
6. Leaves keeled	*C. tropophylla*
5. Megasporophyll apical spine similar to lateral spines	
8. Microsporophyll with distinct terminal spine	
9. Leaves >1 m long, leaflets 14–20 mm wide	
10. Leaves to 360 cm long, leaflets 100–160, megasporophyll blade with 6–9 teeth, seeds 35–40 mm long	*C. tanqingii*
10. Leaves to 240 cm long, leaflets 80–110, megasporophyll blade with 15–25 teeth, seeds 18–27 mm long	*C. chevalieri*
9. Leaves < 1 m long, leaflets 10–15 mm wide	*C. hoabinhensis*
8. Microsporophyll without a terminal spine	
11. Petiolar spines 4–10 mm long	
12. Leaflets 10–13 mm wide, narrowed and tapered at base	*C. balansae*
12. Leaflets 13–21 mm wide	
13. Leaflets abruptly rounded at the base, petiole short	*C. dolichophylla*
13. Leaflets tapered at base, petiole long	*C. aculeata*
11. Petiolar spines 1–3 mm long	
14. 5–10 leaves in crown, leaflets 10–14 mm wide	*C. brachycantha*
14. 1-5 leaves in crown, leaflets >15 mm wide	
15. Leaves with >100 leaflets, male cones 25 cm long, 7 cm or more diam.	*C. collina*
15. Leaves with <110 leaflets, male cones 20 cm long, 6 cm or less diam.	
16. Megasporophyll blade teeth 40–100 mm long	*C. fugax*
16. Megasporophyll blade teeth 15–25 mm long	*C. simplicipinna*
4. Male cones firm, sarcotesta with fibrous layer, sclerotesta plain	
17. Leaflets densely and persistently tomentose, reducing gradually to spines toward leaf base	*C. siamensis*
17. Leaflets loosing tomentum with maturity, leaflets not reducing gradually to spines toward leaf base	
18. Male cones to 20 cm long, 5 cm diam., plants without trunks, leaves strongly keeled	*C. lindstromii*
18. Male cones 25–60 cm long, 5–30 cm diam., at least some plants with trunks, leaves flat or slightly keeled	
19. Male cones >16.5 cm diam., microsporophylls with sharp spine 17–40 mm long	*C. pectinata*

19. Male cones <15 cm diam., microsporophylls with spines 16 mm long or less	
20. Petiole 10–20 cm long, 10–20% of leaf length, spinescent for 80–100% of length	
21. Leaflets 6–8 mm wide, seedling leaflets 5–7 mm wide	*C. elongata*
21. Leaflets 7–10 mm wide, seedling leaflets 7–10 mm wide	*C. pachypoda*
20. Petiole 20 cm or longer, 15–50% of leaf length, unarmed or spinescent for <70% of length	
22. Microsporophyll apical spine < 6 mm long	*C. clivicola*
22. Microsporophyll apical spine > 8 mm long	*C. condaoensis*
2. Megasporophylls not pectinate	
23. Megasporophyll lateral spines distinct, leaves 60–90 cm long, seeds with fibrous sarcotesta but no spongy layer	*C. inermis*
23. Megasporophyll lateral spines not distinct, leaves 150–250 cm long, seeds with no fibrous sarcotesta but spongy layer present	*C. litoralis*
1. Leaflets divided	
24. Leaflets pinnately divided	*C. multipinnata*
24. Leaflets dichotomous	
25. Male cones >6 cm diam., 20 cm long, median microsporophylls rounded	*C. bifida*
25. Male cones up to 5 cm diam., 15 cm long, median microsporophylls with a fine apical spine	*C. micholitzii*

CHAPTER 4: DESCRIPTIONS OF THE SPECIES

The 25 *Cycas* species that are known to occur naturally in Vietnam, together with two Chinese species which probably also occur in the country, are described species-by-species in this Chapter. Complex scientific terminology has been kept to a minimum; where used the terms are consistent with the glossary published by Osborne & Walters (2004). Etymologies are taken from Stearn's Botanical Latin (1992), while author abbreviations follow Brummit & Powell (1992). All place names are given in English spellings; a schedule listing the corresponding Vietnamese names is provided in Appendix 1. The species are grouped according to the sections of the genus in which they occur. The sections are defined as follows:

SECTION *STANGERIOIDES* Smitinand

This section is defined by its soft male cones and soft microsporophylls, pectinate megasporophylls, glabrous ovules and seeds with a sarcotesta that is not fibrous and a wrinkled sclerotesta. Fifteen species occur in Vietnam, mostly in the north. The type species is *Cycas micholitzii* Dyer. Included in this treatment are two Chinese species, *Cycas segmentifida* D. Y. Wang & C. Y. Deng and *C. diannanensis* Z. T. Guan & G. D. Tao; because of their proximity to the Chinese-Vietnamese border it is probable that these also occur in Vietnam.

SECTION *INDOSINENSES* J. Schust.

This section is defined by a combination of glabrous ovules, deeply pectinate megasporophylls and seeds with a fibrous sarcotesta. Seven species occur in Vietnam, mostly in the south. The type species is *Cycas siamensis* Miq.

SECTION *CYCAS*

This section is defined by its glabrous ovules and non-pectinate megasporophylls. Three subsections are recognised but only two occur in Vietnam, subsections *Cycas* and *Rumphiae*.

- Subsection *Cycas* has narrow megasporophyll blades, seeds with a fibrous sarcotesta and without a spongy flotation layer. The sole Vietnamese representative is *Cycas inermis* Lour.

Subsection *Rumphiae* K. D. Hill is uniquely defined by the presence of a spongy layer in the seed. The type species is *Cycas rumphii* Miq. and the sole Vietnamese representative is *C. litoralis* K. D. Hill.

[See also the Key to the sections (Chapter 2, Table 2) and the section and subsection assignments (Chapter 2, Table 3)].

1. *Cycas collina* K. D. Hill, H. T. Nguyen & Phan K. Loc

PUBLICATION

K. D. Hill, Hiêp T. Nguyen & Phan K. Loc. 2004. The Botanical Review 70(2): 142–144.

SYNONYMS

Cycas collina was previously included in *C. simplicipinna* (Smitinand) K. D. Hill.

ETYMOLOGY

From the Latin *collinus*, pertaining to hills, relating to the high ridge country in northern Vietnam where this species occurs.

DISCOVERY

This Vietnamese endemic, like *Cycas fugax* (which follows), was previously referred to as *C. simplicipinna*. The holotype (*S. L. Yang 777 & H. T. Nguyen*) was collected on 5 April 1996 between Hat Lot and Bac Yen villages, Mai Son district, Son La Province, and is filed at the herbarium of the National Center for Natural Sciences and Technology, Hanoi (HN). The species description was published in 2004.

Distribution and habitat: *Cycas collina* is somewhat restricted in its distribution, occurring principally in the mountainous regions of Son La Province in northern Vietnam, generally above 400 or 500 m altitude, in evergreen or partly deciduous forests or woodlands, or bamboo thickets, on steep slopes of mountain ridges. The substrate varies from red clay soils on limestone to sandy loams over sediments.

This species is also expected to occur in similar terrain, perhaps at lower elevations, in Laos. Furthermore, a population of dwarf cycads in the Xishuangbanna region of southern Yunnan, China, previously referred to as *Cycas siamensis* and *C. simplicipinna*, may fall within the present concept of *C. collina*.

DESCRIPTION, VEGETATIVE PARTS

Stems of *Cycas collina* are usually subterranean or procumbent, 15–80 cm long, 8–13 cm in diameter and hold 2–5 leaves in the crown. Leaves are bright to deep green, moderately glossy, with an orange tomentum which is lost as the leaf expands; they are 140–330 cm long and flat (not keeled) in section (opposing leaflets at 150–180° across the rachis). Leaves have 110–180 leaflets and the rachis ends consistently with a pair of leaflets. The petiole is 70–90 cm long (40–55% of total leaf length), glabrous and spinescent for 90–100% of its length. Basal leaflets are 200–260 mm long and are abruptly replaced by petiolar spines. Median leaflets are simple, strongly discolorous, 280–450 mm long, 17–23 mm wide, angled forward (inserted at 80–85° to the rachis), decurrent for 5–10 mm, narrowed to 2.5–4 mm at the base, 15–25 mm apart on the rachis, flat in section, with margins flat or wavy, with the apex softly pointed and with the midrib raised above but not below. Cataphylls are 40–80 mm long, narrowly triangular, soft in texture and lightly hairy.

Cycas collina **– distribution**

FIGURE 21 (a). *Cycas collina*. Plant on steep terrain, Phang Pass, near Son La City, Son La Province. Photo: Ken Hill.

DESCRIPTION, REPRODUCTIVE PARTS

Male cones of *Cycas collina* are solitary, spindle-shaped, cream coloured, 20–25 cm long and 9–11 cm in diameter. Microsporophylls have a soft blade that is not dorsiventrally thickened; they are 12–14 mm long, 7.5–10 mm wide, have a fertile zone 8–10 mm long and a sterile apex that is 3–6 mm long, deflexed and without an apical spine. Megasporophylls are 10–14 cm long, brown-tomentose and have 2–4 glabrous ovules. The megasporophyll blade is ovate, 60 mm long by 30 mm wide, deeply pectinate with 10–14 soft spines about 30 mm long and 2 mm wide; the apical spine is not distinct from the lateral spines. Seeds are ovoid, 25–27 mm long and 18–21 mm wide. The sarcotesta is yellow, not powdery, and non-fibrous. The sclerotesta is slightly warty and there is no internal spongy layer.

DISTINGUISHING FEATURES

Cycas collina falls within a 3-species group that Hill et al. (2004) refer to as the *Cycas simplicipinna* complex, also including *C. fugax*. This group shares a subterranean habit and has reduced megasporophylls lacking a distinct apical spine. Within the group, *C. collina* is distinguished by its few leaves, widely spaced leaflets and relatively large male cones with rounded microsporophylls lacking a terminal spine.

CONSERVATION STATUS

Although *Cycas collina* does not occur in dense populations, the species is widespread. Despite its habitat being continually reduced by agricultural demands, many populations remain and there is no immediate threat of extinction. The total population size is estimated at 2500–10 000, extending over an area of 200 km^2 and the current IUCN Red List status assignment is VU (vulnerable) (Hill et al. 2003).

HYBRIDS AND INTERGRADES

A population of cycads west of the town of Moc Chau in Son La Province differs from the typical *Cycas collina* in having more leaves, leaflets with slender bases, short to medium petioles and small seeds. This is thought to be an intergrade with *C. dolichophylla*, which is abundant at lower altitudes to the east.

USEFUL LITERATURE

Hill et al. (2004) pp. 142–144
Jones (2002) p. 157
Whitelock (2002) p. 103.

FIGURE 21 (b). *Cycas collina.* Leaf detail, from a plant in cultivation, Nong Nooch Tropical Garden, Thailand. Photo: Anders Lindström.

FIGURE 21 (c). A male cone emerging at the stem apex, plant in cultivation, Yen Chau district, Son La Province. Photo: Ken Hill.

FIGURE 21 (d). A female cone emerging amongst cataphylls at the stem apex, plant in cultivation, Thuan Chau town, Son La Province. Photo: Ken Hill.

2. *Cycas fugax* K. D. Hill, H. T. Nguyen & Phan K. Loc

PUBLICATION

K. D. Hill, Hiêp T. Nguyen & Phan K. Loc. 2004. The Botanical Review 70(2): 145–146, fig. 3.

SYNONYMS

As with the previous species, *Cycas fugax* was previously included in *C. simplicipinna* (Smitinand) K. D. Hill.

ETYMOLOGY

From the Latin *fugax*, fleeting or ephemeral, alluding to the near extinction of this species before it was recognised.

DISCOVERY

This Vietnamese endemic, like the preceding *Cycas collina*, was previously referred to *C. simplicipinna*. Specimens were collected in January 2000 at Lam village, Tram Than municipality, Phu Ninh district, Phu Tho Province and the holotype (*T. H. Nguyen 4112 & K . D. Hill*) is filed at the herbarium of the National Center for Natural Sciences and Technology, Hanoi (HN), with an isotype at the Royal Botanic Gardens, Sydney, Australia (NSW). The species description was published in 2004.

DISTRIBUTION AND HABITAT

Cycas fugax is known only from Phu Tho Province in northern Vietnam, occurring at about 200 m in altitude. The habitat was originally closed evergreen forest on low hills but this has been almost completely cleared for tea, *Eucalyptus* and *Acacia* plantations. The substrate is an orange-brown alluvial loam.

DESCRIPTION, VEGETATIVE PARTS:

Cycas fugax has a subterranean flask-shaped stem, about 18 cm long, 15 cm in diameter at the base and 10 cm at the apex, which may be larger and branched in older specimens. Stems hold 1–3 leaves in the crown. Leaves are bright to deep green, moderately glossy, with an orange tomentum which is lost as the leaf expands; they are 280–380 cm long and flat (not keeled) in section (opposing leaflets at 160–180° across the rachis). Leaves have 80–100 leaflets and the rachis ends consistently with a pair of leaflets. The petiole is 150–220 cm long (50–60% of total leaf length), glabrous and spinescent for 90–100% of its length. Basal leaflets are 120–300 mm long and are abruptly replaced by petiolar spines. Median leaflets are simple, strongly discolorous, 400–500 mm long, 18–27 mm wide, angled forward (inserted at 70–85° to the rachis), decurrent for 5–10 mm, narrowed to 2–3 mm at the base, 30–40 mm apart on the rachis, flat in section, with margins flat or wavy, with the apex softly pointed and with the midrib raised above but not below. Cataphylls are 30–50 mm long, narrowly triangular, soft in texture and lightly hairy.

***Cycas fugax* – distribution**

DESCRIPTION, REPRODUCTIVE PARTS

Male cones of *Cycas fugax* are solitary, spindle-shaped, cream in colour, 10–12 cm long and 2.5–4 cm in diameter. Microsporophylls have a soft blade that is not dorsiventrally thickened; they are 12–14 mm long, 7.5–10 mm wide, have a fertile zone 8–10 mm long and a sterile apex that is 3–6 mm long, deflexed and with a slender, closely-appressed apical spine 2–4 mm long. Megasporophylls are 20–25 cm long, brown–tomentose and have 2–4 glabrous ovules. The megasporophyll blade is ovate, 110–130 mm long by 35–40 mm wide, deeply pectinate with 10–14 soft spines 50–90 mm long and 3 mm wide; the apical spine is not distinct from the lateral spines. Seeds are ovoid, 25–27 mm long and 18–21 mm wide. The sarcotesta is yellow, not powdery, and non-fibrous. The sclerotesta is warty and there is no internal spongy layer.

FIGURE 22. *Cycas fugax.* (a) Line drawing by Lesley Elkan. Bar a = 20 cm, b = 5 cm, c = 1 cm. Reproduced from the species description by permission.

FIGURE 22 (c). *Cycas fugax*. Leaf detail, plant in cultivation, Tram Than town. Photo: Piet Vorster.

Left – FIGURE 22 (b). A plant in cultivation, Tram Than town, Phu Tho Province, said to have been collected nearby. Photo: Ken Hill.

FIGURE 22 (d). Male cone emerging at stem apex, plant in cultivation, Tram Than town. Photo: Ken Hill.

FIGURE 22 (e). Cluster of deeply pectinate megasporophylls, plant in cultivation, Tram Than town. Photo: Roy Osborne.

DISTINGUISHING FEATURES

Cycas fugax falls within the 3-species group that Hill et al. (2004) refer to as the *C. simplicipinna* complex, also including *C. collina*. This group shares a subterranean habit and has reduced megasporophylls that lack a clearly distinct apical spine. Within the group, *C. fugax* is distinguished by its few very long leaves, widely spaced leaflets with very narrow tapering bases, small male cones with microsporophylls that have a terminal spine, and megasporophylls with elongated teeth.

CONSERVATION STATUS

As a consequence of land clearing for agriculture, at most only a few representatives remain in the wild. Several plants are found in cultivation in villages near the original habitat and specimens are also known from gardens in Hanoi. This species is the most threatened of the Vietnamese cycads. The total population size is quoted as 250–2 000 occurring over an area of 200 km^2 (Hill et al. 2003), but these figures may be over-estimates.

Cycas fugax is the only species from Vietnam having an IUCN Red List status assignment of CR (critically endangered).

HYBRIDS AND INTERGRADES

Not known.

USEFUL LITERATURE

Hill et al. (2004) pp. 145–146
Jones (2002) p. 168
Whitelock (2002) p. 113.

3. *Cycas simplicipinna* (Smitinand) K. D. Hill

PUBLICATION

K. D. Hill. 1995. Proceedings of the Third International Conference on Cycad Biology: 150.

SYNONYMS

Synonyms for *Cycas simplicipinna* include *C. micholitzii* var. *simplicipinna* Smitinand (1971), *C. balansae* Warb. (Chen & Wang 1995) and *Epicycas tonkinensis* (L. Linden & Rodigas) de Laub. (in part, de Laubenfels & Adema 1998). The name *C. immersa* Craib was misapplied to this taxon by Suvatabandhu (1961).

ETYMOLOGY

From the Latin roots *simplex* and *pinna*, meaning simple leaflets, being in contrast to the divided leaflets of *Cycas micholitzii* and allied species.

DISCOVERY

In 1958 and 1960, Thai botanist Professor Tem Smitinand, Director of the Royal Forest Herbarium in Bangkok, collected specimens of a small cycad in Chiang Mai Province in northern Thailand. Believing this to be a form of *Cycas micholitzii*, he published the species as *C. micholitzii* var. *simplicipinna* in 1971. Ken Hill subsequently upgraded the taxon to species rank in 1995. The female specimen (*Smitinand 4757*), collected at Doi Suthep, Chiang Mai Province and filed in the herbarium of the Royal Forest Department, Bangkok (BKF), is the nominated lectotype.

DISTRIBUTION AND HABITAT

Cycas simplicipinna occurs in the western parts of Quang Tri Province in Vietnam. However, it is also widely distributed in mountainous areas of northern Thailand (Chiang Mai, Loei, Mae Hong Son and Phrae Provinces), as well as northern Myanmar and Laos (Louang Prabang, Phouiou Yhouan and Nam Minh). Sites are consistently above about 600 m in elevation and usually in fairly high rainfall areas. Specimens

***Cycas simplicipinna* – distribution**

are found in deep shade below tall, closed evergreen forests and plants are usually scattered; dense stands are uncommon.

Occurrences of a dwarf cycad from the Xishuangbanna region southern Yunnan, China, under the name *Cycas simplicipinna*, also referred to as *C. siamensis* by earlier Chinese botanists, have not been fully investigated. It is possible that plants of the candidate population correspond more closely to *C. collina*.

DESCRIPTION, VEGETATIVE PARTS

Stems of *Cycas simplicipinna* are usually subterranean or procumbent, 8–14 cm in diameter and hold only 2–5 leaves in the crown. Leaves are bright to deep green, highly glossy, with an orange tomentum which is lost as the leaf expands; they are 90–250 cm long and flat (not keeled) in section (opposing leaflets at 160–180° across the rachis). Leaves have 35–90 leaflets and the rachis ends consistently with a pair of leaflets. The petiole is 35–140 cm long (40–60% of total leaf length), glabrous and spinescent for 60–100% of its length. Basal leaflets are 120–300 mm long and are abruptly replaced by petiolar spines. Median leaflets are simple, strongly discolorous, 200–600 mm long, 14–20 mm wide, angled forward (inserted at 50–65° to the rachis), decurrent for 5–10 mm, narrowed to 2.5–4 mm at the base, 20–40 mm apart on the rachis, flat in section, with margins flat or undulate, with the apex softly pointed and with the midrib raised above but not below. Cataphylls are 40–50 mm long, narrowly triangular, soft in texture, covered with soft hairs and jointed at the base.

DESCRIPTION, REPRODUCTIVE PARTS

Male cones of *Cycas simplicipinna* are solitary, spindle-shaped, cream in colour, 15–21 cm long and 2.2–4 cm in diameter. Microsporophylls have a soft blade that is not dorsiventrally thickened; they are 12–14 mm long, 7.5–10 mm wide, have a fertile zone 8–10 mm long and a sterile apex that is 3–6 mm long, raised and without an apical spine. Megasporophylls are 7–12 cm long, brown-tomentose and have 2 glabrous ovules. The megasporophyll blade is ovate, 45–55 mm long by 18–35 mm wide, deeply pectinate with 10–14 soft spines 15–25 mm long and 1.5–2 mm wide; the apical spine is not distinct from the lateral spines. Seeds are ovoid, 25–27 mm long and 18–21 mm wide. The sarcotesta is yellow, not powdery, and non-fibrous. The sclerotesta is warty and there is no internal spongy layer.

DISTINGUISHING FEATURES

Cycas simplicipinna falls within a 3-species group that Hill et al. (2004) refer to as the *C. simplicipinna* complex, also including *C. collina* and *C. fugax*. This group shares a subterranean habit and has reduced megasporophylls that lack a clearly distinct apical spine. Within the group, *C. simplicipinna* is distinguished by its sparse crown of large, erect leaves with few, broad and well-spaced leaflets, its small male cones, its small megasporophylls with short spines towards the apex, and its small seeds.

CONSERVATION STATUS

Although *Cycas simplicipinna* does not occur in dense populations, the species is widespread. Although its habitat is continually being reduced by demands for agricultural land, many populations remain across a wide range (from north-central Vietnam to north-western Thailand, Myanmar and Laos), and there is no immediate threat of extinction. The total population is estimated to exceed 10 000, occurring over an area of 1000 km^2, and the current IUCN Red List status assignment is NT (near threatened) (Hill et al. 2003).

HYBRIDS AND INTERGRADES

Not known.

USEFUL LITERATURE

Hiêp & Loc (1999) p. 26
Hill & Yang (1999a) p. 35
Hill et al. (2004) pp. 147–148
Jones (2002) pp. 210–211
Smitinand (1971, 1972)
Wang (1996) pp. 117–121
Whitelock (2002) pp. 148–149.

FIGURE 23 (a). *Cycas simplicipinna*. Plants occur in deep shade in closed evergreen forests such as this locality, north of Mae Hong Song City, Mai Hong Song Province, Thailand. Photo: Ken Hill.

FIGURE 23 (b). *Cycas simplicipinna*. Plant in habitat, with local guide, near Fang, northern Thailand. Photo: Willie Tang.

FIGURE 23 (c). Male cones at the pollen-shedding stage, plant in cultivation, Katherine, Australia. Photo: Joe Perner.

FIGURE 23 (d). Leaf detail, plant in cultivation, Nong Nooch Tropical Garden, Thailand. Photo: Anders Lindström.

FIGURE 23 (e). A few new leaves emerging between a cluster of seed-bearing megasporophylls, plant in cultivation, Katherine, Australia. Photo: Joe Perner.

4. *Cycas balansae* Warb.

PUBLICATION

O. Warburg. 1900. Monsunia 1: 179.

SYNONYMS

The present concept of *Cycas balansae* has been cited under the names *Cycas palmatifida* H. T. Chang, Y. Y. Huang & Y. C. Zhong (1998), *C. shiwandashanica* H. T. Chang & Y. C. Zhong (1995), *C. siamensis* Miq. (Cheng et al. 1975), *C. siamensis* subsp. *balansae* (Warb.) Schuster (1932) and *Epicycas tonkinensis* (Linden & Rodigas) de Laub. (in part, de Laubenfels & Adema 1998).

ETYMOLOGY

Named for the French naturalist, Benedict Balansa (1825–1892), who travelled widely in Asia and who discovered the species in north Vietnam.

DISCOVERY

In March 1899, Benedict Balansa collected parts of a cycad he found growing in a pagoda garden in a village near Hanoi. It was from this material that Otto Warburg prepared the brief species description of *Cycas balansae* in 1900. A specimen deposited in Berlin (B), which would probably constitute the type, was lost in World War II; hence a duplicate specimen (*B. Balansa 4084*), at the Muséum National d'Histoire Naturelle in Paris (P), is the designated isotype. Isotypes are filed at the Conservatoire et Jardin Botaniques de la Ville Genèva, Switzerland (G), both in the main herbarium and also in the de Candolle collection (G-DC), and at the Royal Botanic Gardens, Kew, England (K). *C. balansae* was the second of the currently-valid Vietnamese cycad species to be published in the scientific literature (*C. inermis* being the first).

DISTRIBUTION AND HABITAT

Cycas balansae is known from lowland forests in several provinces in northern Vietnam (Lang Son, Quang Ninh, Thai Nguyen, Vinh Phuc and possibly other Provinces where cultivated plants have been seen). Its distribution extends to the Shiwandashan mountains and other coastal parts of Guangxi and Yunnan Provinces in China. The plant is not found in dense stands, but is locally frequent on moist, sheltered slopes and in valleys in deep shade, in tall, evergreen closed forests, at altitudes below 600 m. Geologically, the habitat is described as being loamy soil over acidic rocks such as schists or granites.

***Cycas balansae* – distribution**

DESCRIPTION, VEGETATIVE PARTS

Stems of *Cycas balansae* are usually subterranean, 12–20 cm in diameter and hold 4–9 leaves in the crown. Leaves are deep green, semi-glossy to highly glossy, with a dark brown tomentum that is lost as the leaf expands. They are 120–260 cm long and slightly keeled to flat in section (opposing leaflets at 160–180° across the rachis). Leaves have 90–160 leaflets and the rachis ends consistently with a pair of leaflets. The petiole is 40–160 cm long (30–40% of total leaf length), pubescent with a dark purple–brown tometum and spinescent for 100% of its length. Basal leaflets are 190–120 mm long and are abruptly replaced by petiolar spines. Median leaflets are simple, strongly discolorous, 220–300 mm long, 10–15 mm wide, angled forward (inserted at 75–85° to the rachis), decurrent for 5–8 mm, narrowed to 3–4 mm at the base, 11–21 mm apart on the rachis, flat in section, with margins flat or slightly recurved, with the apex softly pointed and with the midrib raised above but not below. Cataphylls are 60–70 mm long, narrowly triangular, soft in texture and covered with soft hairs.

FIGURE 24 (a). *Cycas balansae.* Shiwandashan Mountain, south-east Gaungxi, China. *Cycas balansae* is locally frequent on the moist sheltered slopes and valleys. Photo: Ken Hill.

FIGURE 24 (b). *Cycas balansae.* Plant in cultivation in a village of cinnamon growers, 30 km north of Ha Long City, Quang Ninh Province. Photo: Roy Osborne.

FIGURE 24 (c). Leaf detail, plant in cultivation, Nong Nooch Tropical Garden, Thailand. Photo: Anders Lindström.

FIGURE 24 (d). Male cone on a plant in cultivation, Son Duong town, Tuyen Quang Province. Photo: Ken Hill.

FIGURE 24 (e). Cluster of megasporophylls with well-developed seeds, plant in cultivation, Nong Nooch Tropical Garden, Thailand. Photo: Anders Lindström.

DESCRIPTION, REPRODUCTIVE PARTS

Male cones of *Cycas balansae* are solitary, spindle-shaped, yellow in colour, 20–25 cm long and 4–7 cm in diameter. Microsporophylls have a soft blade that is not dorsiventrally thickened; they are 16–30 mm long, 11–14 mm wide, have a fertile zone 14–28 mm long and a sterile apex that is 2–4 mm long, level and without an apical spine. Megasporophylls are 8–12 cm long, brown-tomentose and have 2–4 glabrous ovules. The megasporophyll blade is ovate, 40–60 mm long by 30–55 mm wide, deeply pectinate with 14–24 soft spines 25–40 mm long and 2–3 mm wide; the apical spine is not distinct from the lateral spines. Seeds are ovoid, 25–27 mm long and 20 mm wide.

The sarcotesta is yellow, surface not powdery, non- fibrous and 2 mm thick. The sclerotesta is slighty warty and there is no internal spongy layer.

DISTINGUISHING FEATURES

Cycas balansae is one of the many closely-related taxa within section *Stangerioides* that occur through northern Vietnam, Laos, northern Thailand and southern China. Within the group, *C. balansae* is distinguished by its subterranean habit, its few long leaves with long petioles bearing long, slender spines, and its deep purple-brown tomentum that is often long-lasting. Male cones are indeterminate in size, with rounded microsporophylls lacking an apical spine.

Megasporophylls are reduced, with a few long and slender divisions on the blade. Seeds are small.

CONSERVATION STATUS

Cycas balansae is locally frequent in northern Vietnam and parts of southern China. There is no immediate threat of extinction. The total population size is estimated to be greater than 10 000 individuals, extending over an area of about 400 km^2, and the current IUCN Red List status assignment is NT (near threatened) (Hill et al. 2003).

HYBRIDS AND INTERGRADES

Not known.

USEFUL LITERATURE

Hiêp & Loc (1999) p. 25
Hill et al. (2004) pp. 148–149
Jones (2002) pp. 142–143
Wang (1996) pp. 47–49
Whitelock (2002) pp. 88–89
Xiao & Gong (2006).

5. *Cycas aculeata* K. D. Hill & H. T. Nguyen

PUBLICATION

K. D. Hill, Hiêp T. Nguyen & Phan K. Loc. 2004. The Botanical Review 70(2): 149–150.

SYNONYMS

None.

ETYMOLOGY

From the Latin *aculeatus*, prickly or spine-like, referring to the prominent spines on the petioles of this cycad.

DISCOVERY

Ken Hill and Hiêp Nguyen recognised plants in gardens in Cau Hai village, Thua Thien-Hué Province, as being an unusual and possibly novel cycad. Investigations revealed that these cultivated plants had originally been collected from the southern side of the nearby Hai Van Pass, where Hill and Nguyen discovered a single population of the cycads in January 2000.
The holotype (*H. T. Nguyen 4156 & K. D. Hill*) is deposited at the herbarium of the National Center for Natural Sciences and Technology, Hanoi (HN) with an isotype at the Royal Botanical Gardens, Sydney, Australia (NSW). Female reproductive material was seen for the first time when Hill, Nguyen, Osborne and other members of a cycad ecotour visited the Hai Van Pass locality in 2002. The name *Cycas aculeata* became official with the description published in 2004.

DISTRIBUTION AND HABITAT

Cycas aculeata is presently known only from the single locality on the southern slopes of Hai Van Pass, near the coast, in Da Nang Province.
Plants are locally frequent in that area and it is possible that other populations exist in the surrounding hills. The soil is loam over granite on moderate to steep slopes.
The previous forest habitat was severely affected by herbicide

FIGURE 25 (a). *Cycas aculeata*. Southern part of Hai Van Pass, Da Nang Province, the locality for this species. Photo: Roy Osborne.

FIGURE 25 (b). Plant in dense secondary regrowth at the Hai Van Pass locality, with Li Nan. Photo: Roy Osborne.

FIGURE 25 (c). Leaf detail, plant at Hai Van Pass. Photo: Roy Osborne.

FIGURE 25 (d). Cluster of megasporophylls with well-developed seeds, Hai Van Pass. Photo: Roy Osborne.

spraying during the north-south war in the 1970s. The present secondary vegetation consists of dense bamboo, shrub and grass regrowth.

DESCRIPTION, VEGETATIVE PARTS

Stems of *Cycas aculeata* are subterranean, 15–18 cm in diameter and hold 6–23 leaves in the crown. Leaves are deep green, semi-glossy to highly glossy, with an orange tomentum that is lost as the leaf expands. They are 180–250 cm long and slightly keeled to flat in section (opposing leaflets at 130–170° across the rachis). Leaves have 100–150 leaflets and the rachis ends consistently with a pair of leaflets. The petiole is 60–110 cm long (35–50% of total leaf length), glabrous or pubescent and spinescent for 100% of its length. Basal leaflets are 260–380 mm long and are abruptly replaced by petiolar spines. Median leaflets are simple, strongly discolorous, 350–520 mm long, 13–19 mm wide, angled forward (inserted at 45–60° to the rachis), decurrent for 4–6 mm, narrowed to 2–3 mm at the base, 15–28 mm apart on the rachis, slightly keeled in section, with flat margins, with the apex softly pointed and with the midrib raised above and raised or flat below. Cataphylls are 80–100 mm long, narrowly triangular, soft in texture, covered with soft hairs and long-lasting.

DESCRIPTION, REPRODUCTIVE PARTS

Male cones of *Cycas aculeata* are solitary, narrowly ovoid to spindle-shaped, 15–20 cm long and 4–6 cm in diameter. Microsporophylls have a soft blade that is not dorsiventrally thickened; they are 22–25 mm long, 12–14 mm wide, have a fertile zone 20–22 mm long and a sterile apex that is 2–3 mm long, level and without an apical spine.

[Female reproductive parts were not known when the species description was prepared. On the basis of a single female plant that has now been seen, we suggest that the description closely follows that of *C. balansae*, viz: Megasporophylls are 8–12 cm long, reddish–brown tomentose and have 2–4 glabrous ovules.
The megasporophyll blade is ovate, 40–60 mm long by 30–55 mm wide, deeply pectinate with 14–24 soft spines 25–40 mm long and 2–3 mm wide; the apical spine is not distinct from the lateral spines. Seeds are ovoid, 25–27 mm long and 20 mm wide. The sarcotesta is yellow, not powdery, not fibrous and 2 mm thick. The sclerotesta is slighty warty and there is no internal spongy layer.]

DISTINGUISHING FEATURES

Cycas aculeata is one of the many closely-related section *Stangerioides* taxa that occur through northern Vietnam, Laos, northern Thailand, and Yunnan and Guangxi in southern China. Within the group, *C. aculeata* is distinguished by its subterranean stem, its few long leaves with long petioles bearing long, slender spines, and its light orange-brown tomentum. Male cones are small in size, with rounded microsporophylls that lack an apical spine. This species is most similar to the preceding *C. balansae* but differs from the latter in its more numerous leaves with relatively longer petioles, its light orange-brown tomentum, its longer, wider and slightly keeled leaflets that are narrowed at the base, and its longer cataphylls.

CONSERVATION STATUS

Cycas aculeata is not yet very well known. Populations do not fall within any protected area although the Bach Ma National Park to the north-west of its known localities, and the Son Tra Nature Reserve to the south-east, may be close enough to include this species. The population size is estimated to fall between 250 and 2500, occurring over an area possibly as little as 20 km^2. This species is one of three in Vietnam currently assigned an IUCN Red List status of EN (endangered) (Hill et al. 2003).

Cycas aculeata **– distribution**

HYBRIDS AND INTERGRADES

Not known.

USEFUL LITERATURE

Hill et al. (2004) pp. 149–150
Whitelock (2002) pp. 82–83.

FIGURE 26 (a). Fast flowing tributary of the Xiaoheijiang River, Luchun county, south-east Yunnan Province, China, where *Cycas tanqingii* occurs in closed rainforest. Photo: Ken Hill.

FIGURE 26 (b). *Cycas tanqingii* in habitat, Luchun County, showing leaves with characteristically long petioles. Photo: Ken Hill.

FIGURE 26 (c). Leaf detail, specimen 94710D, in cultivation, Montgomery Botanical Center, Florida, USA. Photo: Jody Haynes.

Right – FIGURE 26 (d). Cluster of megasporophylls at the stem apex, specimen 94710D, Montgomery Botanical Center, Florida, USA. Photo: Jody Haynes.

6. *Cycas tanqingii* D. Yue Wang

PUBLICATION

D. Y. Wang. 1996. Cycads of China: 134–137, plate 29, fig. 18.

SYNONYMS

Chen (1999) lists this taxon as synonymous with *Cycas tonkinensis* (Linden & Rodigas) Rodigas.

ETYMOLOGY

Named for Dr. Tan Qing Chen, Director of the Shenzhen Fairly Lake Botanical Garden, China, in recognition of his efforts promoting ex–situ conservation of cycads.

***Cycas tanqingii* – distribution**

DISCOVERY

While visiting the south-east of China's Yunnan Province in May 1995, Chinese botanist Ding Yue Wang noticed plants of a cycad, similar to *Cycas simplicipinna*, growing in the staff gardens of the forestry bureau in Luchun county. The origin of these plants was traced to the nearby Xiaoheijiang River, where Wang collected female reproductive material. Discussions with Ken Hill led to Wang publishing the species *C. tanqingii* in 1996. The holotype (*D. Y. Wang 5538*) is filed in the herbarium at the Shenzhen Fairy Lake Botanical Garden (SZG) with an isotype at Nanjing Forestry University (NF). At the time of preparation of this text, no herbarium specimens had been collected from any plants in Vietnam.

DISTRIBUTION AND HABITAT

Cycas tanqingii is presently known from a fairly wide area in rugged terrain spanning the drainages of the Xiaoheijiang River, Xiao Hejiang district, Luchun county, in south-east Yunnan Province, China, and on a tributary of the Song Da River near the towns of Pa Nam Cum, Ma Li Chi and Phong Tho in Vietnam's northern Lai Chau Province. Specimens occur in understorey to closed rainforest at altitudes less than 800 m. Plants on the Vietnamese side are poorly known as the area is a sensitive military zone and entry is restricted.

DESCRIPTION, VEGETATIVE PARTS

Stems of *Cycas tanqingii* are stout, generally aerial up to 2 m tall, 25–30 cm in diameter and hold 4–7 leaves in the crown. Leaves are deep green, highly glossy, with an orange tomentum that is lost as the leaf expands. They are 190–360 cm long, flat (not keeled) in section (opposing leaflets at 170–180° across the rachis). Leaves have 100–160 leaflets and the rachis ends consistently with a pair of leaflets. The petiole is 70–190 cm long (40–50% of total leaf length), glabrous and spinescent for 90–100% of its length. Basal leaflets are abruptly replaced by petiolar spines. Median leaflets are simple, strongly discolorous, 300–450 mm long, 15–22 mm wide, angled forward (inserted at 65–85° to the rachis), narrowed to 5–6 mm at the base, 15–27 mm apart on the rachis, flat in section with slightly recurved margins, with the apex bristle-like and with the midrib raised above and below. Cataphylls are 50–80 mm long, narrowly triangular, soft in texture, covered with soft woolly hairs and long-lasting.

DESCRIPTION, REPRODUCTIVE PARTS

Male cones of *Cycas tanqingii* are solitary, spindle-shaped, yellow to orange in colour, about 40 cm long and 5–8 cm in diameter. Microsporophylls have a soft blade that is not dorsiventrally thickened; they are 25–30 mm long, 10–13 mm wide, have a fertile zone 22–28 mm long and a sterile apex that is 2–3 mm long, level to raised, and with a slender apical spine, 1.5–2 mm long, that is appressed and sharply upturned. Megasporophylls are 10–12 cm long, brown-tomentose and have 2 glabrous ovules. The megasporophyll blade is

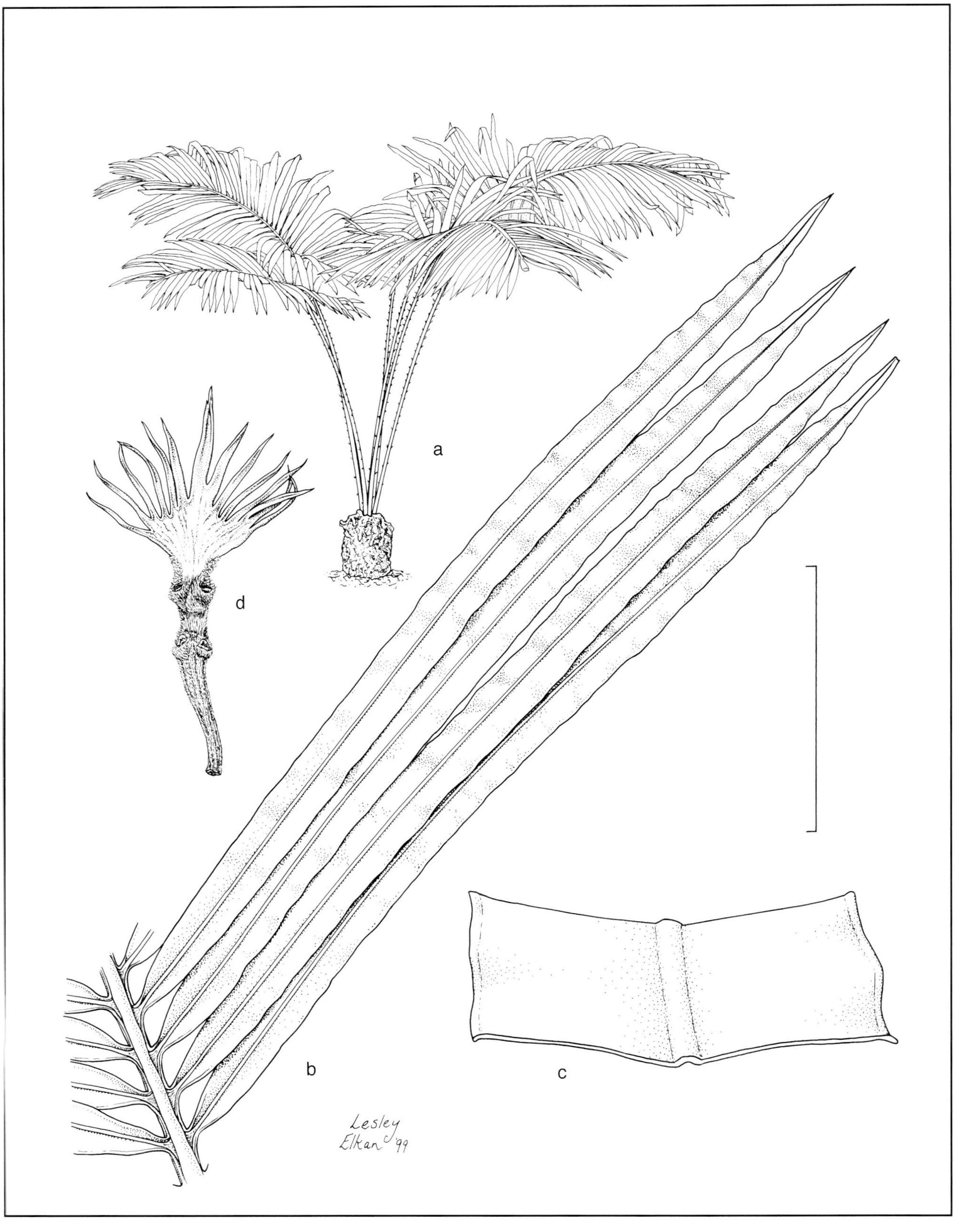

FIGURE 27 (a). ***Cycas chevalieri.*** **Line drawing by Lesley Elkan. Bar a = 50 cm, b = 8 cm, c = 1 cm, d = 6 cm. Reproduced from the species description by permission.**

disc-shaped, 50–55 mm long by 50–65 mm wide, deeply pectinate with 6–9 soft spines 15–40 mm long and 1–3 mm wide; the apical spine is not distinct from the lateral spines, 20–45 mm long by 7–30 mm wide the base. Seeds are almost spherical to ovoid, 35–40 mm long and 30–35 mm wide. The sarcotesta is yellow, not powdery, not fibrous and 2 mm thick. The sclerotesta is warty and there is no internal spongy layer.

DISTINGUISHING FEATURES

Cycas tanqingii falls within group of section *Stangerioides* taxa that occur through northern Vietnam, Laos, northern Thailand, and Yunnan and Guangxi in southern China. Within the group, *C. tanqingii* is distinguished by its robust habit with a stout and generally aerial trunk, its large male cones and its large seeds. This species appears to be most closely related to *C. dolichophylla* but differs in having longer and more widely-spaced leafets that are not broad-based and rounded at the base, and by its longer petioles.

CONSERVATION STATUS

Although *Cycas tanqingii* is not well known, it is clear that the habitat for this species covers a large area in the drainages of the Xiaoheijiang River in China and the Song Da River in Vietnam. The total population size is provisionally estimated at greater than 100 000, extending over an area of at least 100 km^2 and the IUCN Red List status assignment is NT (near threatened) (Hill et al. 2003).

HYBRIDS AND INTERGRADES

Not known.

USEFUL LITERATURE

Chen (1999) pp. 119–120 as *Cycas tonkinensis*
Hill et al. (2004) pp. 150–155
Jones (2002) p. 214
Wang (1996) pp. 134–137
Whitelock (2002) p. 152.

7. *Cycas chevalieri* Leandri

PUBLICATION

J. Leandri. 1931. In: M. H. Lecomte & F. Gagnepain (eds), Flore Générále de l'Indo-Chine 5(10): 1085–1092.

SYNONYMS

This species has been listed in synonymy with *Cycas balansae* Warb. (Chen & Wang 1995, Hiêp & Vidal 1996), and *Epicycas tonkinensis* (Linden & Rodigas) de Laub. (in part, de Laubenfels & Adema 1998).

ETYMOLOGY

Named in honour of Auguste Jean Baptiste Chevalier (1873–1956), a French botanist who spent time in Vietnam and later became Professor of Botany at the Muséum National d'Histoire Naturelle in Paris.

DISCOVERY

This cycad was first collected by Francois Fleury on behalf of Auguste Chevalier from the Nghia Dan area in Vietnam's Nghe An Province. Publication of *Cycas chevalieri* was delayed until 1931 when the French botanist Jacques Leandri described it under its present name in the Flora of Indo-China. Two collections were cited but only the second specimen could ever be located and no further plants were ever found at the original site. The taxon was ignored by Ho in his general enumeration of Vietnamese plants (Ho & Duong 1960, Ho 1991) and it was considered a synonym of *Cycas balansae* by Hiêp & Vidal (1996). However, several new populations were discovered in the period 1994–2000 and specific status for the taxon was clearly warranted. Hill et al. (2002) designated the second specimen (*F. Fleury sub. A. Chevalier 32612*) from the original collection as the lectotype; it is located at the herbarium of the Muséum National d'Histoire Naturelle in Paris (P).

Whitelock (2002) mentions the ethnic name cáy ngen for *C. chevalieri* but we are unable to verify this.

DISTRIBUTION AND HABITAT

Although *Cycas chevalieri* was never rediscovered at the site of its original collection (which is now cleared for cultivation), is presently known from a number of localities in the mountain foothills in Ha Tinh, Nghe An, Quang Binh and Quang Tri Provinces. It is locally abundant in tall, closed evergreen forest on sandy loams over schists and granites. Plants often grow in the form of large multi-

FIGURE 27 (b). *Cycas chevalieri* growing amongst boulders on the banks of the Rao Quan River, near Dak Rong, Quang Tri Province. Photo: Roy Osborne.

FIGURE 27 (c). Leaf detail, plant along Rao Quan River, near Dak Rong, Quang Tri Province. Photo: Piet Vorster.

FIGURE 27 (e). Cluster of deeply pectinate megasporophylls with well-developed glossy seeds, plant near the Rao An River, Ha Tinh Province. Photo: Ken Hill.

Left – FIGURE 27 (d). *Cycas chevalieri* with typical slender aerial trunk, growing amongst boulders alongside the Rao An River, Ha Tinh Province. Photo: Ken Hill.

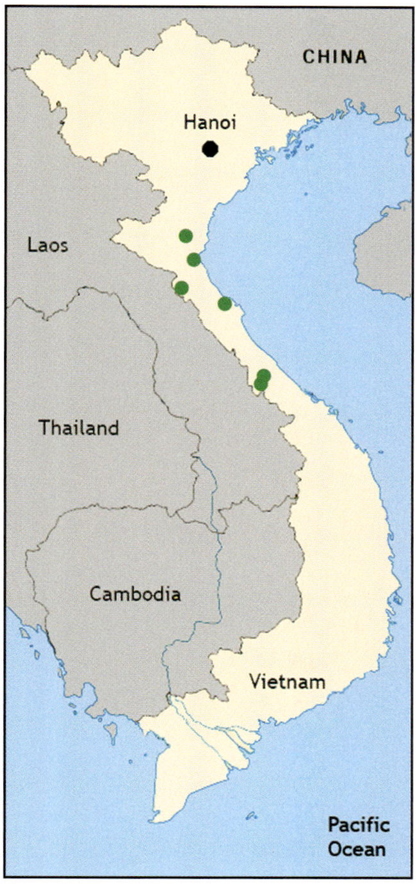

Cycas chevalieri **– distribution**

headed clusters amongst boulders near riverbanks.

DESCRIPTION, VEGETATIVE PARTS

Stems of *Cycas chevalieri* are subterranean to aerial, 8–18 cm in diameter, with 4–15 leaves in the crown. Older plants develop a slender, often decumbent, trunk to 1.2 m long. Leaves are deep green, highly glossy, with an orange or brown tomentum that is lost as the leaf expands. They are 130–240 cm long, slightly keeled to flat in section (opposing leaflets at 150–180° across the rachis). Leaves have 80–110 leaflets and the rachis ends consistently with a pair of leaflets. The petiole is 70–140 cm long (45–60% of total leaf length), glabrous, unarmed or spinescent for 100% of its length. Basal leaflets are 170–240 mm long and are abruptly replaced by petiolar spines. Median leaflets are simple, strongly discolorous, 250–420 mm long, 14–20 mm wide, angled forward (inserted at 50–80° to the rachis), narrowed to 2.5–4 mm at the base, decurrent for 2–6 mm, 15–21 mm apart on the rachis, flat in section with flat or wavy margins, with the apex softly pointed and with the midrib raised above and below. Cataphylls are 50–70 mm long, narrowly triangular, soft in texture and covered with soft hairs.

DESCRIPTION, REPRODUCTIVE PARTS

Male cones of *Cycas chevalieri* are solitary, spindle-shaped, cream or brown in colour, 15–25 cm long and 4–7 cm in diameter. Microsporophylls have a soft blade that is not dorsiventrally thickened; they are 14–17 mm long, 7–10 mm wide and have a raised sterile apex that is without any apical spine. Megasporophylls are 9–13 cm long, brown-tomentose and have 2–4 glabrous ovules. The megasporophyll blade is disc-shaped, 35–55 mm long by 25–50 mm wide, deeply pectinate with 15–25 soft lateral spines 20–35 mm long; the apical spine is only sometimes distinct from the lateral spines, 25–40 mm long by 3–8 mm wide the base. Seeds are ovoid, 18–27 mm long and 15–25 mm wide. The sarcotesta is yellow, not powdery, not fibrous and 1–2 mm thick. The sclerotesta is warty and there is no internal spongy layer.

DISTINGUISHING FEATURES

Cycas chevalieri falls within the group of section *Stangerioides* taxa that occur through northern Vietnam, Laos, northern Thailand and Yunnan and Guangxi in southern China. Within the group, *C. chevalieri* is distinguished by its medium stature, broad leaflets, relatively long petioles that are erect and have short spines that do not continue onto the swollen leaf base. Male cones are small and have microsporophylls with slender apical spines; females have small megasporophylls lacking a clearly defined apical spine. Older plants develop a slender aerial trunk that often falls horizontally with the apex turning vertically. The species is most closely related to *C. hoabinhensis*.

CONSERVATION STATUS

In view of the recent discoveries of large populations in well-preserved forest, this species is considered less at risk than first thought. However, at least some populations have been exploited for the ornamental plant trade in Hanoi. The number of individuals is estimated to exceed 10 000, occurring over an area of 100 km^2 and the proposed IUCN Red List assignment is NT (near threatened) (Hill et al. 2003).

HYBRIDS AND INTERGRADES

Not known.

USEFUL LITERATURE

Hiệp & Loc (1999) p. 26

Hill et al. (2004) pp. 152–153

Jones (2002) pp. 154–155

Whitelock (2002) p. 100.

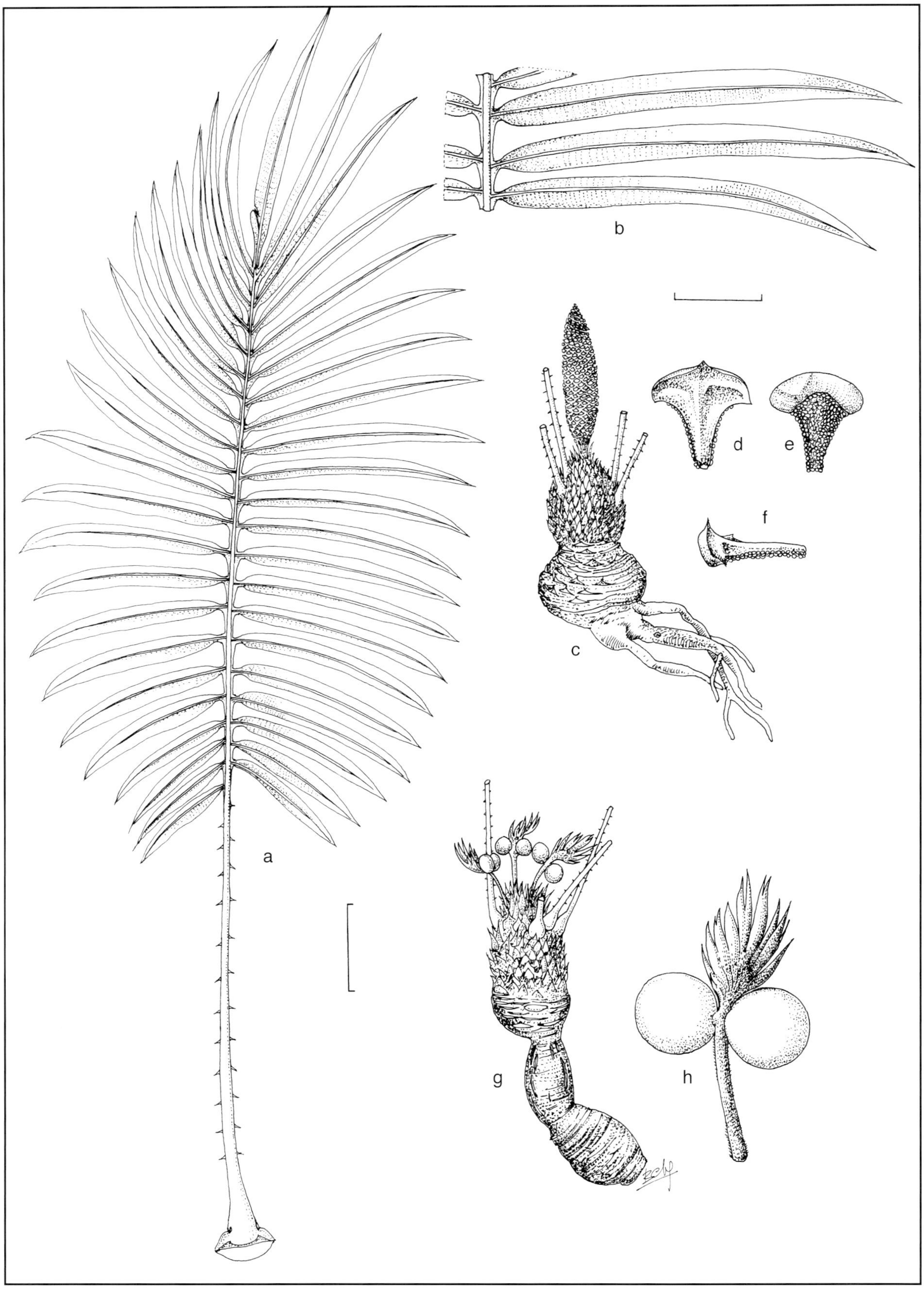

FIGURE 28 (a). ***Cycas hoabinhensis.*** **Line drawing by Bui Xuan Chuong. Bar a,c,g = 8 cm, b = 4 cm, d,e,f = 1 cm, h = 2 cm. Reproduced from the species description by permission.**

8. *Cycas hoabinhensis* Phan K. Loc & H. T. Nguyen

PUBLICATION

K. D. Hill, Hiệp T. Nguyen & Phan K. Loc. 2004. The Botanical Review 70(2): 153–155, fig. 5.

SYNONYMS

This species was included under the concept of *Cycas miquelii* Warb. by Hiệp & Loc (1999) although the authors recognised it as not being identical.

ETYMOLOGY

The epithet for this species recognises its discovery in Hoa Binh Province, northern Vietnam, although it is now known from three other provinces.

***Cycas hoabinhensis* – distribution**

DISCOVERY

Specimens of a small cycad growing to the south of Hanoi have become increasing popular with residents of that city. Fieldwork done over the past 10 years has shown that populations of this plant occur in four provinces to the south of Hanoi. The holotype (*Phan K. Loc P3194a*) was collected in July 1976 (not July 1966 as in Hill et al. 2004) from Chi Ne village, Lac Thuy district in Hoa Binh Province, and filed at the herbarium of the Vietnam National University in Hanoi (HNU), with isotypes deposited at the National Center for Natural Sciences and Technology, Hanoi and the Royal Botanic Gardens, Sydney, Australia (NSW). The species description was published in 2004.

DISTRIBUTION AND HABITAT

Originally found in the Hoa Binh Province, relatively large populations of *Cycas hoabinhensis* have since been discovered in Ha Nam, Ha Tay and Ninh Binh Provinces; all localities are 50–80 km to the south of Hanoi. Plants are locally abundant in sheltered sites on steep limestone outcrops under closed evergreen forest canopy, often in clefts and crevices in bare rock with little or no soil.

DESCRIPTION, VEGETATIVE PARTS

Stems of *Cycas hoabinhensis* are subterranean to aerial up to 0.6 m tall, 5–8 cm in diameter, with 2–10 leaves in the crown. Older plants develop a short trunk that is very slender and often twisted or decumbent. Leaves are bright green, highly glossy, with a white tomentum that is lost as the leaf expands. They are 50–130 cm long, flat (not keeled) in section (opposing leaflets at 170–180° across the rachis). Leaves have 40–100 leaflets and the rachis ends consistently with a pair of leaflets. The petiole is 25–60 cm long (40–60% of total leaf length), glabrous, spinescent for 100% of its length. Basal leaflets are 150–250 mm long and are abruptly replaced by petiolar spines. Median leaflets are simple, strongly discolorous, 200–280 mm long, 11–21 mm wide, angled forward (inserted at 60–80° to the rachis), narrowed to 2.5–5 mm at the base, decurrent for 2–4 mm, 15–20 mm apart on the rachis, flat in section with flat or wavy margins, with the apex softly pointed and with the midrib raised above and below. Cataphylls are 50–70 mm long, narrowly triangular, soft in texture and covered with soft hairs.

DESCRIPTION, REPRODUCTIVE PARTS

Male cones of *Cycas hoabinhensis* are solitary, narrowly ovoid or spindle-shaped, yellow in colour, 10–12 cm long and 5–6 cm in diameter. Microsporophylls have a soft blade that is not dorsiventrally thickened; they are about 22 mm long by 10 mm wide, have a 20 mm long fertile zone and a raised sterile apex 2 mm long. The sterile apex has a distinct or rudimentary apical spine that is pressed against the sporophyll surface, sharply upturned and 4–10 mm long. Megasporophylls are about 7 cm long, brown-tomentose and have 2–4 glabrous ovules. The megasporophyll blade is disc-

FIGURE 28 (b). *Cycas hoabinhensis* occurs on steep limestone outcrops in closed evergreen forests, such as this locality in Cuc Phuong National Park, Ninh Binh Province. Photo: Ken Hill.

FIGURE 28 (c). Potted plants on display at the Huong (Perfume) Pagoda, Ha Tay Province. Photo: Ken Hill.

FIGURE 28 (d). Plant in habitat at Cuc Phuong National Park, Ninh Binh Province. Photo: Ken Hill.

FIGURE 28 (e). Leaf detail, plant in cultivation, Nong Nooch Tropical Garden, Thailand. Photo: Anders Lindström.

shaped, about 30 mm long by 20 mm wide, deeply pectinate with about 10 soft spines about 20 mm long by 2.5 mm wide. The apical spine is not distinct from the lateral spines. Seeds have a yellow sarcotesta that is not powdery and has no fibrous layer. The sclerotesta is warty and there is no internal spongy layer.

DISTINGUISHING FEATURES

Cycas hoabinhensis falls within the group of section *Stangerioides* taxa of subterranean habit that occur in northern Vietnam. Within the group, *C. hoabinhensis* is distinguished by its very small stature, small megasporophylls lacking a distinct apical spine and microsporophylls with a slender apical spine. The species is closely related to *C. chevalieri* and *C. brachycantha*, both of which share its small megasporophyll but are otherwise larger in stature.

CONSERVATION STATUS

The horticultural popularity of this plant in Hanoi, only some 50–80 km to the north, places this species at risk. Although large numbers of plants remain in the less accessible localities, many of the better-known populations have been severely depleted. Populations are preserved within the borders of the Cuc Phuong National Park in Ninh Binh Province, the Chua Huong Tich Nature Reserve in Ha Tay Province and the Thuong Tien Nature Reserve in Hoa Binh Province. The total population size for *Cycas hoabinhensis* is estimated at 2500–10 000, the species occurs over a limited area of about 50 km^2, and the species is one of three in Vietnam that has been assigned an IUCN Red List status of EN (endangered) (Hill et al. 2003).

HYBRIDS AND INTERGRADES

None known.

USEFUL LITERATURE

Hiêp & Loc (1999, as *Cycas miquelii*) pp. 27–28
Hill et al. (2004) pp. 153–155
Jones (2002) p. 171
Osborne et al. (2006)
Whitelock (2002) p. 116.

9. *Cycas brachycantha* K. D. Hill, H. T. Nguyen & Phan K. Loc

PUBLICATION

K. D. Hill, Hiêp T. Nguyen & Phan K. Loc. 2004. The Botanical Review 70(2): 155–157, fig. 6.

SYNONYMS

None.

ETYMOLOGY

The epithet is compounded from the Greek words βραχνς (*brachys*), short, and ακανθα (*acantha*), spine, and refers to the characteristic short spines on the petioles of this plant.

DISCOVERY

This species is another in the series of the smaller cycads of northern Vietnam. It was recognised as a distinct species after visits to the area by Ken Hill and Phan Ke Loc in 1999. The holotype (*H. T. Nguyen 4119 & K. D. Hill*) was collected in January 2000 on Hin Dan Mountain, near Ba Be village in Bac Kan Province, and is filed in the herbarium of the National Center for Natural Sciences and Technology, Hanoi (HN) with an isotype at the Royal Botanic Gardens, Sydney, Australia (NSW). The species description was published in 2004.

Cycas brachycantha is known locally as thiên tuế Ba Bể (Ba Be cycad).

DISTRIBUTION AND HABITAT

Cycas brachycantha is locally frequent in small stands from a limited region in Bac Kan Province in northern Vietnam. Like the preceding *C. hoabinhensis*, this species grows in clefts and crevices of limestone outcrops, with little or no soil, usually on steep slopes under a closed evergreen forest canopy.

DESCRIPTION, VEGETATIVE PARTS

Stems of *Cycas brachycantha* are subterranean to aerial, up to 1 m tall, 9–12 cm in diameter, with 5–10 leaves in the crown. Leaves are deep green, highly glossy, with a white tomentum that is lost as the leaf expands. They are 140–250 cm long, slightly keeled to flat in section (opposing leaflets at 160–180° across the rachis). Leaves have 100–210 leaflets and the rachis ends consistently with a pair of leaflets. The petiole is 50–90 cm long (35–45% of total leaf length), glabrous, spinescent for 90–100% of its length. Basal leaflets are 150–200 mm long and

FIGURE 29 (a). *Cycas brachycantha.* Line drawing by Lesley Elkan. Bar a = 72 cm, b = 10 cm, c = 5 cm. Reproduced from the species description by permission.

are abruptly replaced by petiolar spines. Median leaflets are simple, strongly discolorous, 200–250 mm long, 10–14 mm wide, angled forward (inserted at 70–85° to the rachis), narrowed to 2.5–4 mm at the base, decurrent for 6–8 mm, 10–21 mm apart on the rachis, flat in section with wavy margins, with the apex softly pointed and with the midrib raised above but flat below. Cataphylls are 50–80 mm long, narrowly triangular, sharply pointed and covered with soft hairs.

DESCRIPTION, REPRODUCTIVE PARTS

Male cones of *Cycas brachycantha* are solitary, spindle-shaped, yellow in colour, 12–14 cm long and 3–4 cm in diameter. Microsporophylls have a soft blade that is not dorsiventrally thickened; the sterile apex is not raised and there is no apical spine. Megasporophylls are 8–12 cm long, brown-tomentose and have 2–4 glabrous ovules. The megasporophyll blade is ovate, 25–40 mm long by 15–25 mm wide, deeply pectinate with 14–22 soft spines 15–25 mm long by about 2.5 mm wide. The apical spine is not distinct from the lateral spines. Seeds are ovoid, 25–27 mm long, 20 mm wide, have a yellow sarcotesta with no fibrous layer. The sclerotesta is warty and there is no internal spongy layer.

DISTINGUISHING FEATURES

Cycas brachycantha falls within the group of section *Stangerioides* taxa of mostly subterranean stem that occur in northern Vietnam. Within the group, *C. brachycantha* is distinguished by its medium stature, small male cones on long peduncles having rounded microsporophylls without any distinct apical spine, and reduced megasporophylls also lacking a distinct apical spine. This species is closely related to *C. chevalieri* and *C. hoabinhensis*.

CONSERVATION STATUS

Although restricted to the Bac Kan Province in northern Vietnam, *Cycas brachycantha* is well represented within Ba Be National Park, an area of spectacular beauty. The total population is estimated to exceed 10 000, the area of occurrence is about 50 km^2 and the species is not presently considered to be at risk. The current IUCN Red List status is NT (near threatened) (Hill et al. 2003).

HYBRIDS AND INTERGRADES

Not known.

USEFUL LITERATURE

Hill et al. (2004) pp. 155–157
Jones (2002) pp. 146–147
Whitelock (2002) pp. 92–93.

***Cycas brachycantha* – distribution**

FIGURE 29 (b). Ba Be National Park, Bac Kan Province, an area of spectacular scenery and the locality for this species. Photo: Roy Osborne.

FIGURE 29 (c). *Cycas brachycantha* growing on a steep slope under a closed evergreen forest canopy, Ba Be National Park, Bac Kan Province. Photo: Roy Osborne.

FIGURE 29 (d). Small cluster of megasporophylls at stem apex, Ba Be National Park, Photo: Ken Hill.

FIGURE 29 (e). Male cone at the pollen-shedding stage, plant in cultivation at Nong Nooch Tropical Garden. Photo: Anders Lindström.

FIGURE 29 (f). Leaf detail, plant in cultivation at Nong Nooch Tropical Garden. Photo: Anders Lindström.

FIGURE 29 (g). *Cycas brachycantha* in an ornamental container with Vietnamese boy, Ba Be village. Photo: Jeff Chemnick.

FIGURE 30 (a). *Cycas dolichophylla.* Line drawing by Bui Xuan Chuong. Bar a = 60 cm, b = 24 cm, c = 8 cm, d = 4 cm. Reproduced from the species description by permission.

10. *Cycas dolichophylla* K. D. Hill, H. T. Nguyen & Phan K. Loc

PUBLICATION

K. D. Hill, Hiệp T. Nguyen & Phan K. Loc. 2004. The Botanical Review 70(2): 157–160, fig. 7.

SYNONYMS

None.

ETYMOLOGY

The epithet is derived from the Greek words δολιχος (*dolichos*), long, and φυλλον (*phyllon*), leaf, and refers to the unusually long leaves of this cycad.

DISCOVERY

Although this species is the most widespread and abundant of the cycads of northern Vietnam, occurring also in China, it was recognised as a distinct species only after fieldwork in the 1990s. The holotype (*H. T. Nguyen 2124*) was collected in 1996 from Pu La Mountain, Na Hang district in Tuyen Quang Province, and is filed in the herbarium of the National Center for Natural Sciences and Technology, Hanoi (HN). The species description was published in 2004.

DISTRIBUTION AND HABITAT

Cycas dolichophylla occurs widely throughout the northern parts of Vietnam, and has been recorded from ten Provinces: Bac Kan, Cao Bang, Ha Giang, Lai Chau, Lao Cai, Ninh Binh, Son La, Thai Nguyen, Thanh Hoa and Tuyen Quang. The range extends into Hekou County in the eastern part of China's Yunnan Province. Throughout this broad distribution, the species is locally abundant in sheltered sites in deep shade in closed evergreen forest, although many of the original habitats are now degraded to secondary regrowth and bamboo scrub. Plants grow in loam over limestone, shale, schist or granite substrates.

DESCRIPTION, VEGETATIVE PARTS

Stems of *Cycas dolichophylla* are aerial, up to 1.5 m tall, 18–30 cm in diameter, with 8–40 leaves in the crown. Leaves are bright green to deep green, highly glossy, with an orange tomentum that is lost as the leaf expands. They are 200–450 cm long, flat (not keeled) in section (opposing leaflets at 180° across the rachis). Leaves have 150–270 leaflets and the rachis ends consistently with a pair of leaflets. The petiole is 40–110 cm long (20–35% of total leaf length), glabrous, spinescent for 90–100% of its length. Basal leaflets are 90–240 mm long and are abruptly replaced by petiolar spines. Median leaflets are simple, strongly discolorous, 190–420 mm long, 14–25 mm wide, angled forward (inserted at 60–85° to the rachis), narrowed to 3–6 mm and distinctly rounded at the base, decurrent for 5–8 mm, 16–30 mm apart on the rachis, flat in section with undulate margins, with the apex softly pointed and with the midrib raised above but flat below. Cataphylls are 80–120 mm long, narrowly triangular, soft and covered with soft hairs.

***Cycas dolichophylla* – distribution**

DESCRIPTION, REPRODUCTIVE PARTS

Male cones of *Cycas dolichophylla* are solitary, narrowly ovoid or spindle-shaped, yellow in colour, 35–58 cm long and 8–10 cm in diameter. Microsporophylls have a soft blade that is not dorsiventrally thickened; 30–36 mm long, 9–13 mm wide, with a fertile zone 27–34 mm long; the sterile apex is 2–4 mm long, not raised and the apical spine is absent, or rudimentary, 0–3 mm long and sharply upturned. Megasporophylls are 15–26 cm long, brown-tomentose and have 2–4 glabrous ovules. The megasporophyll blade is disc-shaped, 60–120 mm long by 50–100 mm wide, deeply pectinate with 16–26 soft spines 40–50 mm long by 2–3 mm wide. The apical

FIGURE 30 (b). Evergreen forests on hills near Nam Dat village, Bac Kan Province, one of several localities for *Cycas dolichophylla*. Photo: Ken Hill.

FIGURE 30 (c). *Cycas dolichophylla* in cultivation, Nam Dat village, Photo: Ken Hill.

FIGURE 30 (d). Male cone, plant in cultivation, Nghia Tan village, Ha Giang Province.
Photo: Ken Hill.

Left – FIGURE 30 (e). Leaf detail, plant in cultivation, Nam Dat village. Photo: Piet Vorster.

FIGURE 31 (a). *Cycas micholitzii* in dense forest in central Vietnam. Photo: Si-Lin Yang.

FIGURE 31 (b). *Cycas micholitzii* in cultivation, Ho Chi Minh City. Photo: Ken Hill.

FIGURE 31 (c). Leaf detail, plant in cultivation, Plantation 2000, Brisbane, Australia. Photo: Roy Osborne.

Right – FIGURE 31 (d). Male cone, plant in cultivation, Florida, USA. Photo: Michael Perry.

Left – FIGURE 31 (e). Megasporophylls with well-developed seeds, plant in cultivation, Florida, USA. Photo: Michael Perry.

spine is not distinct from the lateral spines. Seeds are ovoid or flattened ovoid, 40–64 mm long, 30–36 mm wide, have a yellow sarcotesta, 2–4 mm thick, not powdery and without any fibrous layer. The sclerotesta is warty and there is no internal spongy layer.

DISTINGUISHING FEATURES

Cycas dolichophylla is the tenth in our series of section *Stangerioides* taxa of subterranean habit that occur in northern Vietnam. Within the group, *C. dolichophylla* is distinguished by its robust (although usually short) aerial trunk, its longer and more numerous leaves with greater numbers of leaflets, leaflets that are short, broad and rounded at the base, its longer slender petiolar spines, and the larger male cones with rounded microsporophylls lacking an apical projection. Megasporophylls of this species are similar to those of *C. balansae* but larger and more robust.

Hill et al. (2004) believe that *C. dolichophylla* may be part of a natural subgroup of mainly Chinese cycads sharing a stout robust aerial trunk. The subgroup includes *C. hainanensis* C. J. Chen from Hainan, *C. taiwaniana* Carruth. from Guangdong and *C. diannanensis* Z. T. Gaun & G. D. Tao from Yunnan and probably also from Vietnam.

CONSERVATION STATUS

Although *Cycas dolichophylla* is widespread in its occurrence through ten provinces in north Vietnam and its range extension into China, the localities have been subject to significant habitat decline. One population is preserved in the Na Hang Nature Reserve in Tuyen Quang Province while another lies within the Cuc Phuong National Park in Ninh Binh Province. The total plant number is estimated to exceed 10 000, occurring over an area of 500 km^2, and the current IUCN Red List status is VU (vulnerable) (Hill et al. 2003).

HYBRIDS AND INTERGRADES

A population of cycads west of the town of Moc Chau in Son La Province appears to be intermediate between *Cycas dolichophylla* and *C. collina*. In addition, plants seen in cultivation in the Vo Nhai and Yen Son districts in Thai Nguyen Province are morphologically intermediate between *C. dolichophylla* and *C. ferruginea*, while specimens growing in a village west of Thai Nguyen city have characters between those of *C. dolichophylla* and *C. bifida*.

USEFUL LITERATURE

Hill et al. (2004) pp. 157–160
Jones (2002) p. 165
Whitelock (2002) p. 109.

11. *Cycas micholitzii* Dyer

PUBLICATION

W. T. Thiselton-Dyer. 1905. The Gardeners' Chronicle 38: 142–144, figs. 48–49.

SYNONYMS

Synonyms for *Cycas micholitzii* include *Dyerocycas micholitzii* (Dyer) Nakai (1943, as *D. mischolitzi*), and *Epicycas micholitzii* (Dyer) de Laub. (de Laubenfels & Adema 1998).

ETYMOLOGY

Named for William Micholitz (1854–1932), a collector employed by the English nurserymen, Sanders & Sons, and the person who first located this remarkable cycad in Vietnam.

DISCOVERY

It was in 1901 that Micholitz, collecting rare and unusual plants for Sanders and Sons, came across a dwarf species of *Cycas* with dichotomously-divided leaflets in the Amman Highlands near the Laos border. Anticipating the horticultural potential of this species, Micholitz arranged for a large number of specimens to be exported. Most of these were destined for England but a few plants were sent in 1904 to Henry Ridley, Director of the Singapore Botanic Gardens. Ridley sent drawings and specimens to the British botanist Sir William Thiselton-Dyer, who also obtained material from Sanders and Sons in England. Dyer recognised the plant as a species of *Cycas* and first considered it akin to a Chinese species he had described as *C. rumphii* Miq. var. *bifida* (Thiselton-Dyer 1902) (see *C. bifida*). Three years later he clearly separated the Vietnamese taxon as *C. micholitzii* (Thiselton-Dyer 1905), although many subsequent writers have confused the two species. Four sheets in the Kew herbarium (K), from Micholitz's original collection, constitute the holotype. *C. micholitzii* was the third of the currently-valid Vietnamese cycad species to be published in the scientific literature.

Proposals to segregate the species into the genus *Epicycas* de Laub. (de Laubenfels & Adema 1998) or *Dyerocycas* Nakai (1943) have been rejected (Chen et al. 2004).

Micholitz's wild-collected plants initially created considerable interest in England (Prain 1909, Raffill 1912). Male and female plants apparently coned at the Royal Botanic Gardens, Kew. However, most of the imported plants perished with the passage of time and it is possible that the Dublin Botanic Gardens may hold the last known surviving specimen from the original collections (Hill et al. 2004).

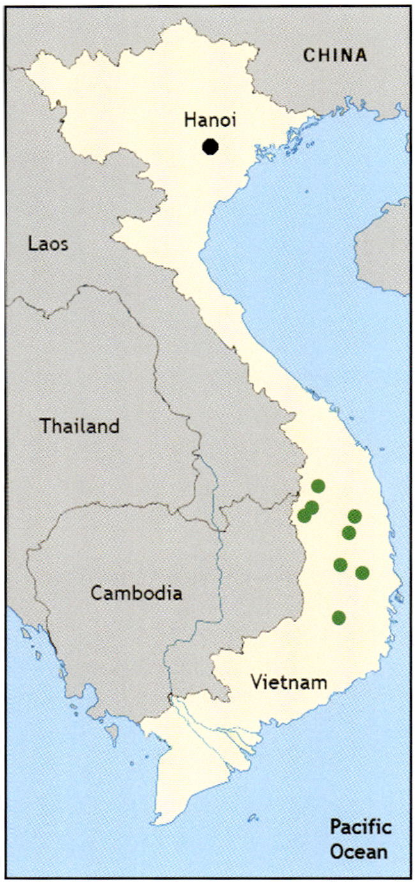

***Cycas micholitzii* – distribution**

DISTRIBUTION AND HABITAT

Cycas micholitzii occurs in the Annam Highlands region of central Vietnam, being recorded from Dak Lak, Gia Lai, Kon Tum and Lam Dong Provinces. Its distribution also extends across the border into adjacent localities in eastern Laos. Plants are found in low, scrubby, but fairly dense woodland and also some parts of the drier monsoon savannah forests. Reports that the plant grows naturally in south-east China (e.g. Chen et al. 1995, Hiêp & Loc 1999, Norstog & Nicholls 1997, Walters et al. 1995; see also Tang et al. 2001) are incorrect and arise from confusion of *C. micholitzii* with *C. bifida* and *C. multipinnata*.

DESCRIPTION, VEGETATIVE PARTS

Stems of *Cycas micholitzii* are subterranean, 10–15 cm in diameter, with 2–6 leaves in the crown. Leaves are bright green, semi-glossy or dull, with a white tomentum that is lost as the leaf expands. They are 110–240 cm long, moderately keeled in section (opposing leaflets at 90–120° across the rachis). Leaves have 18–56 leaflets and the rachis ends consistently with a pair of leaflets. The petiole is 50–105 cm long (40–55% of total leaf length), glabrous, spinescent for 70–95% of its length. Basal leaflets are 230–330 mm long and are abruptly replaced by petiolar spines. Median leaflets are dichotomously branched, strongly discolorous, 230–360 mm long, 11–19 mm wide, angled forward (inserted at 60° to the rachis), narrowed to 2–3 mm at the base, 35–60 mm apart on the rachis, flat in section with flat margins, with the apex softly pointed and with the midrib raised above but flat below. Cataphylls are 60–80 mm long, narrowly triangular, soft and covered with soft hairs.

DESCRIPTION, REPRODUCTIVE PARTS

Male cones of *Cycas micholitzii* are solitary, narrowly ovoid or spindle-shaped, yellow in colour, 15–25 cm long and 3–5 cm in diameter. Microsporophylls have a soft blade that is not dorsiventrally thickened; 13–18 mm long, 7–11 mm wide; the apical spine is prominent or rudimentary, 1–6 mm long and sharply upturned. Megasporophylls have a deeply pectinate blade. Seeds are ovoid, have a sarcotesta without any fibrous layer and a warty sclerotesta with no internal spongy layer.

DISTINGUISHING FEATURES

Cycas micholitzii is one of three Vietnamese species in the genus that have dichotomously divided leaflets as a constant feature. This group also includes *C. bifida* and *C. multipinnata*, both from northern Vietnam and south-east China, and which are treated later in this volume. All these taxa fall within section *Stangerioides* and share a subterranean stem, small, soft male cones, a loose freely-peeling sarcotesta without a fibrous layer, and a warty sclerotesta. Within the group having divided leaflets, *C. micholitzii* is distinguished by its relatively short leaves with dull, comparatively short and narrow leaflets that are dichotomously divided very near to or at the base, and its small male cones with distinct, short, slender spines on the microsporophyll tips.

Conservation status

Cycas micholitzii was relatively unknown in the wild since the time of Micholitz's original collections. In the past two decades, a resurgence of interest in cycads has led to new demands for wild collected plants. This, coupled with habitat clearance for agricultural and forestry, has severely depleted plant numbers in some areas. However, the species is still frequent in many localities, with a total population in excess of 10 000 occurring over an area of about 200 km^2. Populations are preserved in the Ho Lac Nature Reserve, Dak Lak Province, the Mom Ray Nature Reserve, Kon Tum Province and the Deo Ngoan Muc Nature Reserve in Lam Dong Province.
The present IUCN Red List status is VU (vulnerable) (Hill et al. 2003).

Hybrids and intergrades

Not known.

Useful literature

de Laubenfels & Adema (1998, pp. 389–390, as *Epicycas micholitzii*)
Hiêp & Loc (1999) p. 28
Hill et al. (2004) pp. 160–161
Ho & Duong (1960)
Jones (2002) p. 183–184
Leandri (1931)
Norstog & Nicholls (1997) p. 209
Prain (1909)
Smitinand (1971)
Tang et al. (2001)
Wang (1995)
Whitelock (2002) pp. 125–127.

12. *Cycas bifida* (Dyer) K. D. Hill

Publication

K. D. Hill, Hiêp T. Nguyen & Phan K. Loc. 2004. The Botanical Review 70(2): 161–163.

Synonyms

Cycas bifida was first published under the name *C. rumphii* var. *bifida* Dyer (1902). Parts of the concepts of *C. longipetiolula* D. Yue Wang (1996) and *C. multifrondis* D. Yue Wang (1996) are synonyms.

Etymology

From the Latin adjective *bifidus*, two-cleft, in reference to the dichotomously divided leaflets of this cycad.

Discovery

In December 1896, H. B. Morse, customs commissioner, historian and plant collector, gathered leaf material from a dwarf plant with divided leaflets in the Longzhou district of southern Guangxi, China. From a specimen sent to him at the Royal Botanic Gardens, Kew, England, Sir William Thiselton-Dyer recognised it as a cycad and named the Chinese plant as *Cycas rumphii* var. *bifida* in 1902. Dyer noted the similarity of this plant material to that from the Vietnamese taxon he later named as *C. micholitzii* in 1905. Many subsequent writers confused the two species until Hill et al. resolved the situation by elevating the taxon to specific rank as *C. bifida* in 2004. The holotype is represented by two sheets at Kew (K) from the original collection (*Morse 273*).

Distribution and habitat

Cycas bifida occurs in Vietnam's Cao Bang, Lang Son and Tuyen Quang Provinces. Its distribution extends across the border into adjacent localities in Guangxi and Yunnan in China. Plants are abundant in low, scrubby, but fairly dense mixed evergreen and deciduous or bamboo woodland, often on red terra rosa soils on

Cycas bifida **– distribution**

FIGURE 32 (a). Mixed evergreen woodland near Dong Dai village, Tuyen Quang Province, one of the localities for *Cycas bifida*. Photo: Ken Hill.

FIGURE 32 (b). *Cycas bifida*. Leaf detail from a plant near Cao Bang City. Photo: Roy Osborne.

FIGURE 32 (c). (BELOW). Megasporophyll cluster, female plant in cultivation, Shenzhen Fairy Lake Botanical Garden, Guangdong, China. Photo: Ding Yue Wang.

FIGURE 32 (d). Plant in habitat, near Tuc Nga village, Cao Bang Province. Photo: Ken Hill.

FIGURE 32 (e). Megasporophylls with well-developed seed crop, plant in cultivation, California, USA. Photo: Todd Maerowitz.

and around steep karst limestone outcrops or on loamy soils over shales and metasandstones.

Description, vegetative parts

Stems of *Cycas bifida* are subterranean, 10–30 cm in diameter, with 2–5 leaves in the crown. Leaves are bright green, semi-glossy, with a white and orange tomentum that is lost as the leaf expands. They are 200–700 cm long, slightly keeled to flat in section (opposing leaflets at 120–180° across the rachis). Leaves have 40–80 leaflets and the rachis usually ends with a leaflet pair. The petiole is 110–160 cm long (35–50% of total leaf length), glabrous, spinescent for 90–100% of its length. Basal leaflets are abruptly replaced by petiolar spines. Median leaflets are dichotomously branched up to three times, strongly discolorous, 300–600 mm long, 18–25 mm wide, angled forward (inserted at 45–70° to the rachis), decurrent for 10–15 mm, narrowed to 3–4 mm at the base, 60–95 mm apart on the rachis, flat in section with flat margins, with the apex softly pointed and with the midrib raised above but flat below. Cataphylls are 55–90 mm long, narrowly triangular, soft and covered with soft hairs.

Description, reproductive parts

Male cones of *Cycas bifida* are solitary, spindle-shaped, yellow or cream in colour, 35–55 cm long and 6–8 cm in diameter. Microsporophylls have a soft blade that is not dorsiventrally thickened; 17–28 mm long, 13–16 mm wide, with a fertile zone 14–25 mm long and a sterile apex 2–3 mm long, level and without an apical spine. Megasporophylls are 7–11 cm long, brown-tomentose with 2–6 glabrous ovules; the megasporophyll blade is ovate, 35–100 mm long, 25–50 mm wide, deeply pectinate with 12–15 soft lateral spines 18–50 mm long, 2–3 mm wide with an apical spine which may or may not be distinct from the lateral spines, 15–70 mm long by 3–8 mm wide at the base. Seeds are flattened–ovoid, about 25 mm long by 20 mm wide, having a yellow sarcotesta, 2 mm thick, not powdery and without any fibrous layer, and a warty sclerotesta with no internal spongy layer.

Distinguishing features

Cycas bifida is one of three Vietnamese species in the genus that have dichotomously divided leaflets as a constant feature. This group also includes *C. micholitzii* from central Vietnam and *C. multipinnata* from northern Vietnam and south-eastern China. Within the group, *C. bifida* is distinguished by its relatively long leaves with glossy, comparatively long and broad leaflets that are dichotomously divided very near to or at the base. It is most closely related to *C. micholitzii* but has larger male cones with microsporophyll tips that are generally without spines. Despite its original name, the species has no affinity with *C. rumphii*.

Conservation status

Cycas bifida has been severely reduced in numbers from the combined effects of plant collection and land clearance for agriculture and forestry. However, the species is still frequent in many localities in Vietnam and is not considered at risk in the short term. In China, at least one population is protected within the Longgang Nature Reserve in Longzhou, Guangxi. The total plant population is estimated to exceed 10 000, the area of occurrence is about 100 km^2 and the present IUCN status is VU (vulnerable) (Hill et al. 2003).

Hybrids and intergrades

Specimens growing in a village west of Thai Nguyen city have characters between those of *Cycas bifida* and *C. dolichophylla*. In China, intergrades or hybrids are reported between *C. bifida* and *C. multipinnata* (Wang's *C. longipetiolula* being one example of such an intermediate); between *C. bifida* and possibly *C. segmentifida* D. Yue Wang & C. Y. Deng (Wang's *C. multifrondis*); and between *C. bifida* and *C. diannanensis* Z. T. Guan & G. D. Tao.

Useful literature

Chen et al. (1995, as *Cycas micholitzii*) pp. 180–181
Hill et al. (2004) pp. 161–163
Wang (1995)
Wang (1996) pp. 68–76 (as *C. longipetiolula* and *C. micholitzii*) and 80–83 (as *C. multifrondis*)
Whitelock (2002) pp. 90–92.

FIGURE 32 (f). *Cycas bifida*. Male cone, plant near Bai Jue, Guangxi Province, China. Photo: Ken Hill.

FIGURE 33 (a). Closed evergreen forests on steep limestone slope, Manhao, Yunnan Province, China, one of the localitites for *Cycas multipinnata*. Photo: Ken Hill.

FIGURE 33 (b). Chinese botanist inspecting *Cycas multipinnata* in evergreen forest, Yunnan Province. Photo: Si-Lin Yang.

FIGURE 33 (c). *Cycas multipinnata*. Male cone, plant in the Yuan (Red) River catchment, Honghe district of south-east Yunnan, China. Photo: Ding Yue Wang.

FIGURE 33 (d). *Cycas multipinnata*. Leaf detail, plant in habitat in Manhao. Photo: Ken Hill.

13. *Cycas multipinnata* C. J. Chen & S. Y. Yang

PUBLICATION

C. J. Chen & S. Y. Yang. 1994. Acta Phytotaxonomica Sinica 32: 239, 480–481.

SYNONYMS

Cycas multipinnata has been placed in synonymy with *Epicycas multipinnata* (C. J. Chen & S. Y. Yang) de Laub. (de Laubenfels & Adema 1998), *Cycas longipetiolula* D. Yue Wang (1996, in part), and *C. micholitzii* Dyer (Chinese records, in part).

ETYMOLOGY

From the Latin *multi* and *pinna*, referring to the multiple divisions of the primary leaves of this species.

***Cycas multipinnata* – distribution**

DISCOVERY

Although this cycad has been known for several decades, it was only in 1994 that it was described by Professor Chia Jui Chen of the Chinese Academy of Sciences in Beijing, China. The type (*S. Y. Yang & G. X. Mei 9202*) was collected from a plant in cultivation at the Kaiyuan Forestry Station, Yunnan Province, said to have been introduced from Pingbian County, and is filed at the Institute of Botany, Chinese Academy of Sciences, Beijing (PE). Chen (1999) has noted that the type being "from Jainshui" in the protologue is incorrect. Recent botanical exploration has now revealed the presence of a disjunct population of *Cycas multipinnata* in Vietnam.

DISTRIBUTION AND HABITAT

The main populations of *Cycas multipinnata* occur in China, in the Red River (the Yuan River in China and the Song Hong in Vietnam) drainage zone in the Honghe district in the south-east of Yunnan Province. Fieldwork in January 2000 by Hiêp Nguyen and Ken Hill identified a disjunct population of *C. multipinnata* on Chang Re Mountain, in the Yen Binh district of Vietnam's Yen Bai Province. In both the Chinese and Vietnamese localities, the species is found in closed evergreen forests on steep limestone slopes. Much of the original habitat has been cleared or is severely degraded.

The segregation of the species into the genus *Epicycas* de Laub. (de Laubenfels & Adema 1998) has been rejected (Chen et al. 2004).

DESCRIPTION, VEGETATIVE PARTS

Stems of *Cycas multipinnata* are subterranean, 14–25 cm in diameter, bearing only 1–2 leaves in the crown. Leaves are deep green, highly glossy, with a white tomentum that is lost as the leaf expands. They are 200–600 cm long, slightly keeled in section (opposing leaflets at 120–150° across the rachis). Leaves have 14–36 bipinnate leaflets and the rachis consistently ends with a leaflet pair. The petiole is 100–350 cm long (50–65% of total leaf length), glabrous, spinescent for 90–100% of its length. Basal leaflets are 500–720 mm long and are abruptly replaced by petiolar spines. Median leaflets are typically twice dichotomously branched, strongly discolorous, 400–600 mm long, with segments 15–22 mm wide, angled forward (inserted at 45–70° to the rachis), 160–200 mm apart on the rachis, flat in section with flat margins, with the apex softly pointed and with the midrib raised above but flat below. Cataphylls are 40–60 mm long, narrowly triangular, soft, covered with soft hairs and jointed basally.

DESCRIPTION, REPRODUCTIVE PARTS

Male cones of *Cycas bifida* are solitary, spindle-shaped, cream in colour, 25–40 cm long and 6–8 cm in diameter. Microsporophylls have a soft blade that is not dorsiventrally thickened, with a sterile apex that is level or raised, and without an apical spine. Megasporophylls are 11–13 cm long, brown-tomentose with 2–6 glabrous ovules; the megasporophyll blade is ovate, 60–70 mm long, 50–60 mm wide, deeply pectinate with 20–22 soft

FIGURE 34 (a). *Cycas ferruginea.* Plants grow on the exposed surfaces of steep limestone monoliths in north-east Vietnam, such as this example about 50 km west of Lang Son City, Lang Son Province. Photo: Roy Osborne.

FIGURE 34 (b). Plant in rock face habitat, near Long Fu Shan, China. Photo: Phillip Cribb.

FIGURE 34 (c). Emerging male cone, plant in cultivation, Quang Son village, Thai Nguyen Province, said to have been collected from the Huu Lien Nature Reserve. Photo: Ken Hill.

FIGURE 34 (d). Cluster of deeply pectinate megasporophylls, plant in cultivation, Hanoi. Photo: Ken Hill.

lateral spines 30–40 mm long, 1.5–2 mm wide with an apical spine which may or may not be distinct from the lateral spines, 30–40 mm long by 3–4 mm wide at the base. Seeds are flattened-ovoid, about 25 mm long by 21 mm wide, having a yellow sarcotesta, 2 mm thick, not powdery and without any fibrous layer, and a warty sclerotesta with no internal spongy layer.

DISTINGUISHING FEATURES

Cycas multipinnata is the last of three Vietnamese species in the genus that have dichotomously divided leaflets as a constant feature. This group also includes *C. micholitzii* from central Vietnam and *C. bifida* from northern Vietnam and south-eastern China. The species considered closest to *C. multipinnata* is China's *C. debaoensis* Y. C. Zhong & C. J. Chen, from which the former is distinguished by its longer leaves with broader but thinner leaflets that are broadest above the midpoint, and its smallest megasporophylls with shorter teeth towards the blade apex.

CONSERVATION STATUS

Like the preceding species, *Cycas multipinnata* has been severely reduced in numbers from the combined effects of plant collection and land clearance for agriculture and forestry. Although the Vietnamese range has not yet been fully explored; a provisional total population number is 1000–2500 and the area of occurrence is approximately 250 km^2. This species is one of three Vietnamese species with a current IUCN Red List status of EN (endangered) (Hill et al. 2003).

HYBRIDS AND INTERGRADES

Hybrids have been reported in China between *C. multipinnata* and *C. bifida* (Wang's *C. longipetiolula* being one example of such an intermediate).

USEFUL LITERATURE

Chen (1999) pp. 117–118
Chen et al. (1995) p. 188
de Laubenfels & Adema (1998, p. 391, as *Epicycas multipinnata*)
Hill et al. (2004) p. 164
Huang et al. (1999)
Jones (2002) pp. 187–188
Wang (1995)
Wang (1996) pp. 86–90
Whitelock (2002) pp. 128–129.

14. *Cycas ferruginea* F. N. Wei

PUBLICATION

F. N. Wei. 1994. Guihaia 14: 300.

SYNONYMS

Wang (1996) considered *Cycas ferruginea* as synonymous with *C. miquelii* Warb.

ETYMOLOGY

From the Latin *ferrugineus*, rusty-red, in reference to the abundant red tomentum in the crown of this species.

DISCOVERY

This cycad was described in 1994 by the Chinese botanist Fan Nan Wei from a single plant in cultivation at the Guilin Botanical Garden in Guangxi, China. The holotype (*F. N. Wei 2220*) is filed at the Garden's own herbarium (IBK). Specimens from the same plant (*Chen, Zhong & Hill 43*), collected in November 1998, are filed at the Chinese Academy of Sciences, Beijing (PE) and at the Royal Botanic Gardens, Sydney, Australia (NSW). A population of wild plants has been located in western Guangxi, China, and field studies in the late 1990s revealed a significant number of plants in Vietnam's Lang Son and Thai Nguyen Provinces. These plants do not differ significantly from the cultivated Chinese type specimen.

DISTRIBUTION AND HABITAT

Apart from its occurrence in China, large stands of *Cycas ferruginea* are present in Vietnam in the Huu Lien Nature Reserve, Huu Lung district, Lang Son Province and to the west in the Dong Hy district of Thai Nguyen Province. This species, like the two which follow (*C. sexseminifera* and *C. tropophylla*), is restricted in its Vietnamese habitat to a belt of steep limestone monoliths in the north-east. Plants grow on exposed surfaces with roots in only a little organic detritus in rock crevices.

DESCRIPTION, VEGETATIVE PARTS

Stems of *Cycas ferruginea* are subterranean to aerial, to 1.2 m tall, 12–18 cm in diameter, bearing 10–25 leaves in the crown. Leaves are deep green or grey-green, highly glossy or semi-glossy, with a white, orange or brown tomentum on new growth, this shading later to a characteristic rusty-red tomentum that is lost as the leaf expands but persists below and in the crown. They are 130–210 cm long, slightly keeled or flat in section (opposing leaflets at 150–180° across the rachis). Leaves have 120–200 leaflets and the rachis usually ends with a leaflet pair. The petiole is 35–80 cm long (25–40% of total leaf length), glabrous or pubescent, spinescent for 30–100% of its length. Basal leaflets are 80–190 mm long and are abruptly replaced by petiolar spines. Median leaflets are simple, often falcate, strongly discolorous, 190–380 mm long, 8–13 mm wide, angled slightly forward (inserted at 70–90° to the rachis), decurrent on the rachis, slightly keeled in section with recurved margins, with a sharply spined apex, and with the midrib raised above but flat below. Cataphylls are 90–120 mm long, narrowly triangular, sharply-pointed and long-lasting.

***Cycas ferruginea* – distribution**

DESCRIPTION, REPRODUCTIVE PARTS

Male cones of *Cycas ferruginea* are solitary, spindle-shaped, orange in colour, 25–45 cm long and 4–6 cm in diameter. Microsporophylls have a soft blade that is not dorsiventrally thickened, 16–20 mm long, 9–13 mm wide, with a fertile zone 15–18 mm long and a sterile apex that is raised, 3–5 mm long, and with a rudimentary, sharply upturned apical spine 0–3 mm long or absent. Megasporophylls are 11–19 cm long, brown-tomentose with 2–6 glabrous ovules; the megasporophyll blade is disc-shaped, 70–110 mm long, 50–80 mm wide, deeply pectinate with 28–42 soft lateral spines 18–40 mm long, 1–3 mm wide with an apical spine distinct from the lateral spines, 15–50 mm long by 3–8 mm wide at the base. Seeds are ovoid, 19–24 mm long by 15–21 mm wide, having a yellow sarcotesta, 1–2 mm thick, not powdery and without any fibrous layer, and a warty sclerotesta with no internal spongy layer.

DISTINGUISHING FEATURES

Cycas ferruginea is considered to be closest to *C. sexseminifera* (which follows), but is clearly distinguished by the dense rusty-red tomentum on new growth, its longer leaves, and its longer, often falcate leaflets that have recurved margins. It is also separated on account of its generally more robust habit and its longer and broader megasporophylls with a distinct, broad apical spine.

CONSERVATION STATUS

A significant population of *Cycas ferruginea* is protected within the Huu Lien Nature Reserve, Huu Lung district in Lang Son Province. The total population size for the species is estimated to exceed 100 000, the area of occurrence is about 150 km^2. The species is not considered to be at risk in the short term and the current IUCN Red List assignment is NT (near threatened) (Hill et al. 2003).

HYBRIDS AND INTERGRADES

Plants in cultivation in the Vo Nhai and Yen Son districts in Thai Nguyen Province may represent a hybrid between *Cycas ferruginea* and *C. dolichophylla*.

USEFUL LITERATURE

Chen (1999) pp. 125–126
Hiêp & Loc (1999) pp. 26–27
Hill et al. (2004) pp. 165–166
Jones (2002) pp. 167–168
Whitelock (2002) p. 112.

15. *Cycas sexseminifera* F. N. Wei

PUBLICATION

F. N. Wei. 1996. Guihaia 16: 1–2.

SYNONYMS

This species was previously considered as *Cycas miquelii* Warb. by numerous authors (Chen 1999, Chen & Stevenson 1999, de Laubenfels & Adema 1998, Hiêp & Loc 1999, Wang 1996). Other synonyms are *C. acuminatissima* H. T. Chang, Y. C. Zhong & Z. F. Lu (1998), *C. brevipinnata* H. T. Chang, Y. Y. Huang, Y. C. Zhong (1998), *C. inermis* Lour. (Metcalf 1942), *C. longisporophylla* F. N. Wei (1997), *C. pectinata* Buch.-Ham. (Cheng et al. 1975, Zhou et al. 1990, both in part), *C. septemsperma* H. T. Chang, Y. Y. Huang & H. Z. Zheng (1998), and *C. spiniformis* J. Y. Liang (Wei 1997).

ETYMOLOGY

The epithet is compounded from the Latin words *sex*, six, and *seminifer*, bearing, in reference to the supposedly characteristic presence of six ovules on each megasporophyll.

DISCOVERY

After many years of confusion, principally with the concept of *Cycas miquelii* Warb. (see synonyms, below), this taxon was formally described in 1996 by the Chinese botanist Fan Nan Wei from a plant in cultivation at the Guilin Botanical Garden in Guangxi, China, said to have been introduced from Longzhou County. The holotype (*F. N. Wei 2223*) is filed at the Garden's own herbarium (IBK).

DISTRIBUTION AND HABITAT

Cycas sexseminifera is widespread in the southern and central parts of China's Guangxi Province and adjacent parts of Cao Bang Province in north-eastern Vietnam. Disjunct Vietnamese populations have also been found in Ninh Binh and Thanh Hoa Provinces, south of Hanoi. The species is locally abundant in its areas of occurrence. Like the closely related species, *C. ferruginea* and *C. tropophylla*, it grows in crevices of rugged karst limestone monoliths, often on vertical faces with roots in only a little organic detritus.

DESCRIPTION, VEGETATIVE PARTS

Stems of *Cycas sexseminifera* are subterranean to aerial, 0–1 m tall, 8–20 cm in diameter, and bear 4–20 leaves in the crown.

Leaves are deep green, semi-glossy, with a brown tomentum that is lost as the leaf expands. They are 50–120 cm long, slightly keeled to flat in section (opposing leaflets at 150–180° across the rachis). Leaves have 80–200 leaflets and the rachis consistently ends with a leaflet pair. The petiole is 10–40 cm long (20–35% of total leaf length), glabrous, unarmed or spinescent for 5–100% of its length. Basal leaflets are 10–80 mm long and are abruptly replaced by petiolar spines. Median leaflets are simple, strongly discolorous, 130–240 mm long, 6–10 mm wide, angled slightly forward (inserted at 70–85° to the rachis), decurrent for 2–6 mm on the rachis, slightly keeled in section with flat or slightly recurved margins, with a sharply-pointed apex and with the midrib raised above and below. Cataphylls are 30–50 mm long, narrowly triangular, soft, covered with soft hairs and long-lasting.

DESCRIPTION, REPRODUCTIVE PARTS

Male cones of *Cycas sexseminifera* are solitary, narrowly ovoid to spindle-shaped, yellow in colour, 15–30 cm long and 5–8 cm in diameter. Microsporophylls have a soft blade that is not dorsiventrally thickened, 20–30 mm long, 12–30 mm wide, with a fertile zone 10–28 mm long and a sterile apex that is raised, 2–4 mm long, and

***Cycas sexseminifera* – distribution**

FIGURE 35 (a). *Cycas sexseminifera* grows on rugged karst limestone monoliths such as in the area near Lung Trang village, Cao Bang Province. Photo: Ken Hill.

FIGURE 35 (b). Plant in cultivation, Nong Nooch Tropical Garden, Thailand. Photo: Anders Lindström.

FIGURE 35 (c). Leaf detail, plant in cultivation, Nong Nooch Tropical Garden, Thailand. Photo: Anders Lindström.

with a rudimentary, sharply upturned apical spine upto 3 mm long or absent. Megasporophylls are 8–12 cm long, brown-tomentose with 2–6 glabrous ovules; the megasporophyll blade is disc-shaped, 35–55 mm long, 30–50 mm wide, deeply pectinate with 16–28 soft lateral spines 10–25 mm long, with an apical spine distinct from the lateral spines, 30–40 mm long. Seeds are ovoid, 20–28 mm long by 18–25 mm wide, having a yellow sarcotesta, not powdery and without any fibrous layer, and a smooth to warty sclerotesta with no internal spongy layer.

DISTINGUISHING FEATURES

Cycas sexseminifera is distinguished by its dwarf habit; its short stem bears short, flat leaves with short, narrow, flat, rather stiff and sharply-pointed leaflets; its small but broad megasporophyll blade with a broad apical spine distinct from the lateral spines. Petioles often lack spines but this is not always the case. *C. sexseminifera* is closely allied to the previous species, *C. ferruginea*.

CONSERVATION STATUS

There is considerable horticultural demand for *Cycas sexseminifera* both in China and in Vietnam. However, the population size of this species is estimated to exceed 10 000, in an area of occurrence of about 500 km^2 and the species is not considered to be at risk in the short term. The current IUCN Red List assignment is NT (near threatened) (Hill et al. 2003).

HYBRIDS AND INTERGRADES

Not known.

USEFUL LITERATURE

Chen (1999, as *Cycas miquelii*) p. 126
Hiệp & Loc (1999, as *C. miquelii*) pp. 27–28
Hill et al. (2004) pp. 166–168
Wang (1996, as *C. miquelii*), pp. 76–80
Whitelock (2002) pp. 145–146.

FIGURE 35 (d). Male cone of *Cycas sexseminifera*, plant at Shanyu, China. Photo: Ken Hill.

FIGURE 35 (e). Cluster of megasporophylls emerging amongst apical cataphylls, plant at Ping Xian, China. Photo: Ken Hill.

FIGURE 36 (a). *Cycas tropophylla.* **Line drawing by Bui Xuan Chuong. Bar a = 20 cm, b = 6 cm, c = 2.66 cm, d = 2 cm. Reproduced from the species description by permission.**

16. *Cycas tropophylla* K. D. Hill & Phan K. Loc

PUBLICATION

K. D. Hill, Hiêp T. Nguyen & Phan K. Loc. 2004. The Botanical Review 70(2): 168–170.

SYNONYMS

None.

ETYMOLOGY

The epithet is compounded from the Greek words τροπις (*tropis*), a boat keel, and φυλλον (*phyllon*), leaf, in reference to the distinctively keeled leaf profile of this species.

DISCOVERY

Botanical explorations in 1994, 1996 and 1999 by Si-Lin Yang, Hiêp Nguyen, Phan Loc, Ken Hill, Nguyen Van Dzu and Anders Lindström led to the formal description in 2004 of a cycad in the Ha Long Bay area in north-eastern Vietnam. The holotype (*K. D. Hill 5020, Phan K. Loc & N. V. Dzu*), collected on 10 March 1999, is on file at the herbarium of the National Center for Natural Sciences and Technology, Hanoi (HN); isotypes are deposited at the Royal Botanic Gardens, Sydney, Australia (NSW), at Kew Herbarium in England (K) and at the New York Botanical Garden (NY). *Cycas tropophylla* is the last of the section *Stangerioides* cycads known from Vietnam to be dealt with in this volume. The two species which follow (*C. segmentifida* and *C. diannanensis*) are Chinese species and expected to be found in Vietnam after further exploration along the China-Vietnam border.

Known locally as thiên tuế Hạ Long (Ha Long cycad), *C. tropophylla* is sometimes grown locally as a potted specimen plant (Hiêp & Kiew 2000).

DISTRIBUTION AND HABITAT

Cycas tropophylla occurs over a large sea area but a fairly limited land area that includes many of the smaller islands in Ha Long Bay (Quang Ninh Province) together with the major landmass of Cat Ba Island (Haiphong Municipality) and several limestone bluffs on the adjacent mainland areas. Osborne (2005) considers this as the most beautiful cycad habitat in the world, a view widely shared by those who have visited the area. Like the two preceding taxa, *C. ferruginea* and *C. sexseminifera*, this species grows on near-vertical limestone cliff faces with roots in little or no soil in rock crevices.

***Cycas tropophylla* – distribution**

FIGURE 36 (b). *Cycas tropophylla*. Potted male specimen in display in Cam Pha Town, Quang Ninh Province. Photo: Ken Hill.

FIGURE 36 (c). Ha Long Bay, an area of great beauty with more than 2000 limestone monolith islands, and one of the areas where *Cycas tropophylla* occurs. Photo: Roy Osborne.

FIGURE 36 (d). *Cycas tropophylla* growing on a near-vertical cliff face of an island in Ha Long Bay. Photo: Willie Tang.

FIGURE 36 (e). Leaf detail showing the keeled nature of the foliage, plant in cultivation, Plantation 2000, Brisbane, Australia. Photo: Roy Osborne.

FIGURE 36 (f). Male cone, from a plant in cultivation in Hanoi. Photo: Ken Hill.

FIGURE 36 (g). Cluster of megasporophylls on a plant in cultivation near Ha Long City, Quang Ninh Province. Photo: Willie Tang.

DESCRIPTION, VEGETATIVE PARTS

Stems of *Cycas tropophylla* are subterranean to aerial, to 1 m tall, 12–20 cm in diameter, bearing 10–40 leaves in the crown. Leaves are deep green, semi-glossy, with a brown tomentum that is lost as the leaf expands. They are 130–210 cm long, strongly keeled* in section (opposing leaflets at 90–120° across the rachis). Leaves have 120–200 leaflets and the rachis consistently ends with a leaflet pair. The petiole is 35–80 cm long (25–35% of total leaf length), glabrous, and spinescent for 30–100% of its length. Basal leaflets are 80–190 mm long and are abruptly replaced by petiolar spines. Median leaflets are simple, strongly discolorous, 190–270 mm long, 8–13 mm wide, angled slightly forward (inserted at 70–90° to the rachis), decurrent for 2–5 mm on the rachis, narrowed to 2.5–4 mm at the base, 10–16 mm apart on the rachis, slightly keeled in section with slightly recurved margins, with the apex aristate, spinescent, and with the midrib raised above and below. Cataphylls are 100–120 mm long, narrowly triangular, soft in texture, covered with soft hairs and long-lasting.

[*This differs from the data given in Hill (2004), which was based on cultivated plants. There are also differences to the data given in Whitelock (2002) that was apparently based on somewhat stressed plants growing in deep shade on Cat Ba Island and which were not wholly typical.]

DESCRIPTION, REPRODUCTIVE PARTS

Male cones of *Cycas tropophylla* are solitary, narrowly ovoid to spindle-shaped, yellow in colour, 25–45 cm long and 4–6 cm in diameter. Microsporophylls have a soft blade that is not dorsiventrally thickened, 16–20 mm long, 9–13 mm wide, with a fertile zone 15–18 mm long and a 3–5 mm sterile apex that is raised, 3–5 mm long, and with a rudimentary apical spine upto 2 mm long or absent. Megasporophylls are 11–16 cm long, brown-tomentose with 2–6 glabrous ovules; the megasporophyll blade is disc-shaped, 70–120 mm long, 55–80 mm wide, deeply pectinate with 24–42 soft lateral spines 18–35 mm long, 1–2 mm wide with an apical spine distinct from the lateral spines, 15–21 mm long by 4–8 mm wide at the base. Seeds are ovoid, 20–28 mm long by 18–25 mm wide, having a yellow sarcotesta, 1–2 mm thick, not powdery and without any fibrous layer, and a smooth to warty sclerotesta with no internal spongy layer.

DISTINGUISHING FEATURES

Cycas tropophylla is distinguished from *C. ferruginea* by its keeled leaves, its shorter flat leaflets and an absence of any significant dark tomentum on the new growth. It is separated from *C. sexseminifera* again by its keeled leaves, its generally more robust habit, its larger and broader megasporophylls with a distinct, broad apical spine.

CONSERVATION STATUS

Significant populations of this species are protected within the Cat Ba National Park in Hai Phong Municipality and the Ha Long Bay Nature Reserve in Quang Ninh Province. The population size of *Cycas tropophylla* is estimated to exceed 10 000 although the area of occurrence is only about 50 km^2. The species is not considered to be at risk in the short term and the current IUCN Red List assignment is NT (near threatened) (Hill et al. 2003).

HYBRIDS AND INTERGRADES

None known.

USEFUL LITERATURE

Hill et al. (2004) pp. 168–170
Jones (2002) p. 217
Osborne (2005)
Whitelock (2002) pp. 154–155.

FIGURE 37 (a). *Cycas segmentifida*. Line drawing by Shi-Jing Zhang. Reproduced from the species description by permission.

17. *Cycas segmentifida* D. Yue Wang & C. Y. Deng

PUBLICATION

D. Y. Wang & C. Y. Deng. 1995. Encephalartos 43: 11–14, figs 1–2.

SYNONYMS

Synonyms for *Cycas segmentifida* include *C. longiconifera* H. T. Chang, Y. C. Zhong & Y. Y. Huang (1998), C. multifida H. T. Chang & Y. C. Zhong (1977) and *C. xilingensis* H. T. Chang & Y. C. Zhong (1997). Reference to the present concept of *C. segmentifida* has also been made under the names *C. hainanensis* C. J. Chen var. *latifolia* C. Y. Deng, *C. guizhouensis* K. M. Lan & R. F. Zou, and *C. guizhouensis* var. *latifolia* C. Y. Deng, while the names *C. balansae* Warb. and *C. siamensis* Miq. have been misapplied to this taxon (see Wang 1996).

***Cycas segmentifida* – distribution**

ETYMOLOGY

The epithet for this species is compounded from the Latin words *segmentum*, segment, and *fidus*, divided, and refers to the fine dichotomous branching of the spines on the margins of the megasporophyll blade.

DISCOVERY

Vigorous interest by Chinese cycad botanists in the 1990s resulted in the naming of several 'new' species from the south-western parts of China. Some of these have been internationally accepted while others appear to be synonyms or fail to meet criteria for recognition. Like the situation in Vietnam, many of these plants were first noticed in cultivation and follow-up fieldwork often led to the discovery of wild populations. *Cycas segmentifida* is a case in point. Ding Yue Wang and Han Peng noticed an unusual female cycad in cultivation in the South China Botanical Garden, Guangzhou Province and collected a specimen on 16 May 1994 (*D. Y. Wang & H. Peng 2967*) which serves as the type for the species. In October 1994, Wang met with Cao Yi Deng, of the Forest Research Institute of Southwest Guizhou, who had been investigating wild populations of the same plant; together they published the species description in 1995. The holotype is deposited in the herbarium of the Shenzhen Fairy Lake Botanic Garden, Guangdong, China (SZG), with isotypes at Chinese herbaria in Guizhou (GZAC, XIN) and Nanjing (NF). Additional isotypes are filed at Fairchild Tropical Garden, Miami (FTG) and Sydney, Australia (NSW).

DISTRIBUTION AND HABITAT

Cycas segmentifida occurs in China's western Gaungxi Province, with range extensions into southern Guizhou and eastern Yunnan Provinces.
It is believed that populations of this species may occur in adjacent sites in Vietnam's Ha Giang and Cao Bang Provinces but this is yet to be verified. Plants in the Chinese localities occur at low altitudes on steep slopes, with little soil cover, over substrates from limestone to shale and schist. The vegetation varies from closed evergreen forest to mixed deciduous and evergreen woodland but much of this is now degraded to grassland or scrubby regrowth.

DESCRIPTION, VEGETATIVE PARTS

Stems of *Cycas segmentifida* are subterranean or aerial to 0.5 m tall, 10–23 cm in diameter, and bear 8–25 leaves in the crown. Leaves are deep green, highly glossy, with an orange tomentum that is lost as the leaf expands. They are 120–280 cm long, flat (not keeled) in section (opposing leaflets at 180° across the rachis). Leaves have 90–190 leaflets and the rachis consistently ends with a leaflet pair. The petiole is 40–140 cm long (30–50% of total leaf length), glabrous and spinescent for 90–100% of its length. Basal leaflets are 70–290 mm long and are abruptly replaced by petiolar spines. Median leaflets are simple, strongly discolorous, 120–420 mm long, 9–20 mm wide,

FIGURE 37 (b). Degraded grassland and scrubby regrowth on steep sites at Long Lo Tun, near Xilin City, Guangxi, China, one of several localities for *Cycas segmentifida*. Photo: Ken Hill.

FIGURE 37 (c). Plant in habitat, Yan Ron, near Tianlin city, Guangxi, China. Photo: Ken Hill.

FIGURE 37 (d). *Cycas segmentifida.* Leaf detail, plant in cultivation, Plantation 2000, Brisbane Australia. Photo: Roy Osborne.

FIGURE 37 (e). Male plant with old and new cones, from a plant in cultivation at the South China Botanical Garden, Guangdong, China. Photo: Ding Yue Wang.

Left – FIGURE 37 (f). Megasporophylls with fully-developed seeds, plant at Long Lo Tun locality. Photo: Ken Hill.

angled forward (inserted at 50–80° to the rachis), decurrent for 4–7 mm, narrowed to 3–7 mm at the base, 11–24 mm apart on the rachis, flat in section with flat or wavy margins, with a sharp apex (not spinescent), and with the midrib raised above and below. Cataphylls are 70–100 mm long, narrowly triangular, soft, covered with soft hairs and long-lasting.

DESCRIPTION, REPRODUCTIVE PARTS

Male cones of *Cycas segmentifida* are solitary, spindle-shaped, yellow in colour, 30–60 cm long and 5–12 cm in diameter. Microsporophylls have a soft blade that is not dorsiventrally thickened, 20–30 mm long, 15–18 mm wide, with a fertile zone 17–27 mm long and a sterile apex about 3 mm long, level, and without any apical spine. Megasporophylls are 16–25 cm long, yellow or brown-tomentose, with 4–8 glabrous ovules; the megasporophyll blade is disc-shaped, 80–130 mm long, 50–120 mm wide, deeply pectinate, with 16–44 finely-divided, soft, lateral spines 20–70 mm long, 2–3 mm wide, with an apical spine distinct from the lateral spines, 25–125 mm long by 8–20 mm wide at the base. Seeds are near-spherical, 28–35 mm long and 24–30 mm wide, with a yellow sarcotesta 1.5 mm thick, not powdery, with no fibrous layer, a warty sclerotesta, and no internal spongy layer.

DISTINGUISHING FEATURES

Cycas segmentifida is distinguished amongst the Chinese section *Stangerioides* taxa by its small seeds, broad leaflets that are not twisted on the rachis, with flat or wavy margins, and a large megasporophyll apex with a broad apical spine. The Latin diagnosis given by Wang & Deng (1995) states '… *macrosporophylla lamina terminali segmenti fidi plerumque tenuis*' emphasising the finely-divided lateral spines on the megasporophyll blade for which the species is named. *C. segmentifida* is most closely related to *C. szechuanensis* C. Y. Cheng, W. C. Cheng & L. K. Fu from Fujian and Guangdong Provinces, and C. guizhouensis K. M. Lan & R. F. Zou from Guangxi, Guizhou and Yunnan Provinces.

CONSERVATION STATUS

Until fairly recently, *Cycas segmentifida* was widespread through China's western Gaungxi Province, with range extensions into southern Guizhou and eastern Yunnan Provinces. However, populations have been severely reduced through deforestation and the present number of mature plants in China is estimated at less than 10 000 in an area of occupancy of about 500 km^2. The long leaves and small stems of this species make it less desirable for horticulture than other Chinese cycads and there is little threat of widespread demand for wild specimens. The current IUCN Red List status assignment of VU (vulnerable) (Hill et al. 2003), may be revised if sizeable populations are found in Vietnam.

HYBRIDS AND INTERGRADES

The species named as *Cycas multifrondis* D. Yue Wang (1996) may represent a hybrid between *Cycas bifida* and *C. segmentifida.*

USEFUL LITERATURE

Chen & Stevenson (1999)
Chen (1999) pp. 120–121
Deng (1999)
Hill et al. (2004) pp. 188–189
Jones (2002) pp. 205–206
Wang (1996) pp. 108–111
Whitelock (2002) pp. 144–145.

18. *Cycas diannanensis* Z. T. Guan & G. D. Tao

PUBLICATION

Z. T. Guan & G. D. Tao. 1995. Sichuan Forestry Survey & Design 4: 1–2, figs 1–7.

SYNONYMS

Synonyms for *Cycas diannanensis* are *C.* '*parvulus*' (correctly *C. parvula*) S. L. Yang (Wang 1996), and *C. pectinata* var. *manhaoensis* C.J. Chen & P. Yun (1995).

ETYMOLOGY

The epithet for *Cycas diannanensis* refers to the Diannan region, Jianchuan County in south Yunnan Province, China, the type locality for this cycad.

DISCOVERY

As with the preceding *Cycas segmentifida*, the taxon now known as *C. diannanesis* was the result of vigorous interest by Chinese cycad botanists in the 1990s. Its taxonomy has been confusing: Shu Hua Huang discovered a population of this plant in 1994 in a rainforest at Manhao, near Gejiu City in southern Yunnan

FIGURE 38 (a). *Cycas diannanensis.* Plants occur at altitudes of 600–1800 m, usually on steep slopes; much of the original cloud forest habitat is degraded to grassland or scrubby secondary regrowth. Near Manhoa, Yunnan, China. Photo: Ken Hill.

FIGURE 38 (b). Plant in cultivation, Shenzhen Fairy Lake Botanic Garden, Guangdong, China. Photo: Ken Hill

FIGURE 38 (c). Male cone, plant in cultivation, Shiping, China. Photo: Ken Hill.

Left – FIGURE 38 (d). Leaf detail, Nong Nooch Tropical Garden, Thailand. Photo: Anders Lindström.

Province, but identified it as *C. siamensis*. Guo Da Tao and Ding Yue Wang separately collected material from the same site in 1995 and recognised it as a probable new species. Chia Jui Chen and Yun (1995) published it under the name *C. pectinata* subsp. *manhaoensis* while Guan & Tao (1995) described it as *C. diannanensis* at the same time, depositing the holotype (*G. D. Tao 95014*) at The Sichuan Forestry Exploration and Design Institute (SFEDI) with an isotype at Xishuangbanna (HITBC). Chen (1999) regarded the taxon as synonymous with *C. taiwaniana* Carruth. However, Hill et al. (2004) recognised *C. diannanensis* as being widespread in Yunnan, and also including a population from near Mengdian village in Honghe County which Si Lin Yang had previously named as *C. 'parvulus'* (Wang 1996).

DISTRIBUTION AND HABITAT

Cycas diannanensis occurs widely in the central and eastern parts of China's Yunnan Province, along the Red River (Honghe) valley between Hekou and Shuangbai. It is believed that populations of this species occur in adjacent sites in Vietnam's Lao Cai Province, where the Red River is known as the Song Hong. Plants in the Chinese localities occur at altitudes of 600–1800 m, usually on steep slopes, over substrates from limestone to shale and schist. The original vegetation was closed evergreen cloud forest but much of this is now degraded to grassland or scrubby secondary regrowth.

[Hiệp and Loc's (1999) report that *C. diannanensis* occurs in Vietnam's Quang Ninh and Tuyen Quang Provinces was based on a misidentification.]

DESCRIPTION, VEGETATIVE PARTS

Stems of *Cycas diannanensis* are subterranean or aerial to 3 m tall, 25–35 cm in diameter, and bear 12–50 leaves in the crown. Leaves are bright green or deep green, highly glossy, with a white or orange tomentum that is lost as the leaf expands. They are 140–330 cm long, slightly keeled or flat in section (opposing leaflets at 160–180° across the rachis). Leaves have 160–300 leaflets and the rachis either ends with a spine 1–11mm long or with a leaflet pair. The petiole is 45–100 cm long (20–30% of total leaf length), glabrous and spinescent for 90–100% of its length. Basal leaflets are 30–240 mm long and are abruptly replaced by petiolar spines. Median leaflets are simple, strongly discolorous, 130–320 mm long, 9–16 mm wide, angled forward (inserted at 50–80° to the rachis), decurrent for 3–7 mm, narrowed to 2.5–4 mm at the base, 10–20 mm apart on the rachis, flat in section with slightly recurved margins, with a sharp apex (not spinescent), and with the midrib raised above and below. Cataphylls are 120–200 mm long, linear, sharply-pointed, covered with silky hairs or lacking any tomentum, and long-lasting.

DESCRIPTION, REPRODUCTIVE PARTS

Male cones of *Cycas diannanensis* are solitary, narrowly ovoid in shape, yellow in colour, 50–65 cm long and 9–13 cm in diameter. Microsporophylls have a soft blade that is not dorsiventrally thickened, 35–60 mm long, 15–25 mm wide, with a fertile zone 30–50 mm long and a sterile apex 4–10 mm long, raised, and with a rudimentary sharply upturned apical spine 0–2 mm long. Megasporophylls are 16–24 cm long, grey-tomentose, with 4–8 glabrous ovules; the megasporophyll blade is disc-shaped or ovate, 65–140 mm long, 55–140 mm wide, deeply pectinate, with 30–52 soft, lateral spines 10–50 mm long, 1.5–4 mm wide, with an apical spine distinct from the lateral spines, 10–45 mm long by 7–15 mm wide at the base. Seeds are ovoid, 28–40 mm long and 20–32 mm wide, with a yellow sarcotesta 3–4 mm thick, not powdery, with no fibrous layer, a warty sclerotesta, and no internal spongy layer.

DISTINGUISHING FEATURES

Cycas diannanensis is distinguished amongst the Chinese section *Stangerioides* taxa by its very long

Cycas diannanensis **– distribution**

and sharply-pointed cataphylls. It is more robust than the closely-related *C. segmentifida*, but also has small seeds, broad leaflets that are not twisted on the rachis, with flat or wavy margins, and a large megasporophyll apex with a broad apical spine. Hill et al. (2004) consider that *C. diannanensis* may be part of a natural subgroup of mainly Chinese cycads sharing a stout robust aerial trunk; the subgroup includes *C. hainanensis* C. J. Chen from Hainan, *C. taiwaniana* Carruth. from Guangdong and *C. dolichophylla* K. D. Hill, H. T. Nguyen & Phan K. Loc from Vietnam.

CONSERVATION STATUS

As with *Cycas segmentifida*, populations of *C. diannanensis* have been reduced through deforestation. The present number of mature plants in China is estimated at 50 000 in an area of occupancy of about 50 km^2. The current IUCN Red List status assignment is VU (vulnerable) (Hill et al. 2003), but this may be revised if sizeable populations are found in Vietnam.

HYBRIDS AND INTERGRADES

A hybrid population between *C. bifida* and *C. diannanensis* is reported from Yunnan (Hill et al. 2004).

USEFUL LITERATURE

Hiêp & Loc (1999) p. 26
Hill et al. (2004) pp. 189–190
Jones (2002) pp. 164–165
Wang (1996) pp. 49–51, and pp. 93–96 as *C. parvulus*
Whitelock (2002) p. 108.

FIGURE 38 (e). ***Cycas diannanensis.*** **Cluster of megasporophylls emerging at the stem apex, Manhao, Yunnan Province, China. Photo: Ken Hill.**

FIGURE 39 (a). ***Cycas pectinata.*** **Partially degraded habitat for the species near Sa Huynh town, Quang Ngai Province, with the South China Sea in the backround. Photo: Roy Osborne.**

FIGURE 39 (b). ***Cycas pectinata.*** **An old plant from Da Lat, Lam Dong Province, 10 m tall and branched. Photo: Si Lin Yang.**

19. *Cycas pectinata* Buch.-Ham.

PUBLICATION

F. Buchanan-Hamilton. 1826. Memoirs of the Wernerian Natural History Society 5(2): 322–323, figs 3 & 5. It is noted that this species is often incorrectly cited as *Cycas pectinata* Griff. (see Pant et al. 1994).

SYNONYMS

Synonyms for *Cycas pectinata* include *C. circinalis* L. subsp. *vera* J. Schust. var. *pectinata* Griff. (Schuster 1932), *C. dilatata* Griff. (Griffith 1854), and *C. jenkinsiana* Griff. (Griffith 1854).

ETYMOLOGY

From the Latin adjective *pectinatus*, comb-like, in reference to the highly pectinate morphology of the megasporophylls.

DISCOVERY

Francis Buchanan-Hamilton was a Scottish surgeon who practised in Bengal, India, from 1795 to 1815, and who also took on the rôle of superintendent of the Calcutta Botanic Garden. His many plant collecting trips lead to his description of *Cycas pectinata* from the hills near Bengal, but material sent to Kew and to Edinburgh did not include a specimen that could serve as the type. There is some controversy as to the legality of the name as Hamilton quoted a validly-published Australia species, *C. angulata* R. Br., as a synonym for *C. pectinata*. However, he did so with a question mark in the text, and therefore his description stands as acceptable in the Rules of Botanical Nomenclature. A redescription of the species by William Griffith in 1854 is thus technically invalid. The matter of a type was recently resolved by Hill et al. (2004) who have designated Kew specimen, *J. D. Hooker & Thomson 6*, collected in Chittagong, India, in 1855, as the neotype; an isoneotype exists in the Paris (P) herbarium.

DISTRIBUTION AND HABITAT

Cycas pectinata is the most widespread of the Asian cycads, occurring in large numbers over a huge area in as many as nine countries (Bangladesh, Bhutan, China, Cambodia, India, Laos, Nepal, Thailand and Vietnam). It is also the most widely distributed cycad in Vietnam, found in 15 provinces (Ba Ria-Vung Tau, Binh Dịnh, Binh Thuan, Da Nang, Dac Lak, Dong Nai, Gia Lai, Khanh Hoa, Kien Giang, Kon Tum, Lam Dong, Ninh Thuan, Phi Yen, Quang Nam and Quang Ngai Provinces). Within this broad distribution, most populations are sparse and occur in medium to tall forest on deep, clay-rich or more fertile soils, usually in moist conditions in moderate to deep shade. Nowadays, many of the plants exist in degraded habitats in exposed situations. Populations are typically found at 500–800 m altitudes, often on limestone substrates, but also on granite or metasediments.

DESCRIPTION, VEGETATIVE PARTS

Stems of *Cycas pectinata* are aerial, up to 12 m tall with branching in older specimens, 14–20 cm in diameter, with a smooth, thin bark, and bear 30–40 leaves in the crown. Leaves are deep green to grey-green, semi-glossy, with a white tomentum that is lost as the leaf expands. They are 140–240 cm long, flat (not keeled) in section (opposing leaflets at 170–180° across the rachis). Leaves have 180–312 leaflets and the rachis consistently ends with a spine 1–46 mm long. The petiole is 30–80 cm long (25–50% of total leaf length), glabrous, spinescent for 30–80% of its length. Basal leaflets are 50–160 mm long and are abruptly replaced by petiolar spines. Median leaflets are simple, strongly discolorous, 200–315 mm long, 7.5–10.5 mm wide, angled forward

***Cycas pectinata* – distribution in Vietnam**

FIGURE 39 (c). *Cycas pectinata* plant near Sa Huynh town, showing the arborescent habitat. Photo: Roy Osborne.

FIGURE 39 (d). Leaf detail, plant near Sa Huynh town. Photo: Piet Vorster.

FIGURE 39 (e). Tight cluster of megasporophylls, after pollination and fertilisation, plant in habitat, Thailand. Photo: Ken Hill.

FIGURE 39 (f). Male cone, plant in habitat on the Thailand-Myanmar border. Photo: Joe Perner.

(inserted at 45–60° to the rachis), decurrent for 4–8 mm, narrowed to 2.5–4 mm at the base, 8–13 mm apart on the rachis, flat in section with slightly recurved margins, with the apex acute, spinescent or not spinescent, and with the midrib raised above and below. Cataphylls are narrowly triangular, soft in texture and covered with soft hairs, and jointed at the base.

DESCRIPTION, REPRODUCTIVE PARTS

Male cones of *Cycas pectinata* are solitary, ovoid in shape, yellow or green in colour, 30–55 cm long and 16–22 cm in diameter. Microsporophylls have a firm blade that is not dorsiventrally thickened, 43–60 mm long, 19–24 mm wide, with a fertile zone 35–57 mm long and a sterile apex that is level (not raised), 3–8 mm long, and with a prominent, sharply upturned apical spine 17–32 mm long. Megasporophylls are 22–30 cm long, grey-tomentose with 2–4 glabrous ovules; the megasporophyll blade is disc-shaped, about 110 mm long, 100–130 mm wide, deeply pectinate with 40–50 soft lateral spines 26–75 mm long, 2–3 mm wide with an apical spine distinct from the lateral spines, 35–75 mm long by 5–12 mm wide at the base. Seeds are flattened-ovoid, 42–45 mm long by 33–45 mm wide, having a yellow sarcotesta, 4–7 mm thick, not powdery and with a fibrous layer, and a smooth sclerotesta with no internal spongy layer.

DISTINGUISHING FEATURES

Cycas pectinata is distinguished from its allies in the *C. pectinata* group by its very large, ovoid, male cones with long narrow microsporophylls with long apical spines. The smooth, thin bark separates *C. pectinata* from many in the group, although this character is also common to *C. clivicola* and *C. elongata*. This species is also the tallest of the Vietnamese cycads with older specimens often branched. A branched specimen in the Mengla Nature Reserve, China, was estimated to reach 13 m in height (Vorster 1996).

CONSERVATION STATUS

Significant populations of this species are protected within the Bu Gia Map and Kon Cha Rang Nature Reserves in Vietnam. The population size of *Cycas pectinata* is believed to exceed 200 000 individuals globally, extending over an area of about 3 000 km^2 and the species is not considered to be at risk in the short term. However, the current IUCN Red List assignment is VU (vulnerable) in view of the widespread and continuing loss of habitat (Hill et al. 2003).

HYBRIDS AND INTERGRADES

None reported.

USEFUL LITERATURE

Chen (1999) p. 127
Hiêp & Loc (1999) p. 31
Hill et al. (2004) pp. 170–173
Hill & Yang (1999a) p. 37
Jones (2002) pp. 194–195
Norstog & Nicholls (1997) p. 206
Pant (1962)
Pant et al. (1994)
Smitinand (1972)
Tang et al. (1997) p. 10
Vorster (1996)
Wang (1996) pp. 97–100
Whitelock (2002) pp. 134–135.

20. *Cycas clivicola* subsp. *lutea* K. D. Hill

PUBLICATION

K. D. Hill & S. L. Yang. 1999b. Brittonia 51: 64, figs 8e, f & i.

SYNONYMS

Epicycas elongata (Leandri) S. L. Yang (in part, de Laubenfels & Adema 1998) was used in synonymy with this species.

ETYMOLOGY

The Latin species epithet, *clivicola*, means 'dwelling on cliffs' while *lutea* refers to the characteristic yellow colour of the subspecies' trunk.

DISCOVERY

Fieldwork by Ken Hill and Poonsak Vatcharacorn in Thailand in 1994 led to the naming of *Cycas clivicola* K. D. Hill and the associated subspecies *lutea* in 1999. The types for both taxa are filed at the herbarium of the Royal Botanic Gardens, Sydney, Australia (NSW), the subspecies voucher being *K. D. Hill & Poonsak Vatcharakorn 4620*. Comparison of this material with vouchers from collections made in

Cambodia in 1870 and 1954 showed that subspecies *lutea* extends into that country; further collections in 1994 revealed that the taxon is also present in southern Vietnam.

DISTRIBUTION AND HABITAT

Cycas clivicola subsp. *lutea* is widespread in south-east Thailand and adjacent coastal Cambodia, and extends marginally into An Giang and Kien Giang Provinces in southern Vietnam*. Like some of the section *Stangerioides* species from northern Vietnam (e.g. *C. ferruginea*, *C. sexseminifera* and *C. tropophylla*), this cycad grows on near-vertical cliff faces of limestone outcrops, with roots in clefts and crevices with little or no soil.

[Hill et al. (2004) include Minh Hai Province (now divided into Ca Mau and Bac Lieu Provinces) in the area of occurrence, but there is no real evidence that the distribution extends much further than the localities immediately adjacent the Cambodian border.]

***Cycas clivicola* subsp. *lutea* – distribution**

DESCRIPTION, VEGETATIVE PARTS

Stems of *Cycas clivicola* subsp. *lutea* are aerial, up to 6 m tall, 12–16 cm in diameter, with a smooth, thin, yellow-coloured bark, and bear 45–100 leaves in the crown. Stems of older plants on cliff faces are often procumbent or J-shaped. Leaves are grey-green initially, bright green later, semi-glossy, with a white tomentum that is lost as the leaf expands. They are 90–160 cm long, slightly keeled to flat in section (opposing leaflets at 150–180° across the rachis). Leaves have 130–200 leaflets and the rachis usually ends with a spine 2–30 mm long. The petiole is 20–45 cm long (20–30% of total leaf length), glabrous, spinescent for 0–70% of its length. Basal leaflets are 50–200 mm long and are abruptly replaced by petiolar spines. Median leaflets are simple, weakly discolorous, 100–260 mm long, 5–11 mm wide, angled forward (inserted at 60–80° to the rachis), decurrent for 3–8 mm, narrowed to 1.5–3.5 mm at the base, 8–14 mm apart on the rachis, flat in section with flat margins, with the apex acute, spinescent, and with the midrib raised above and below. Cataphylls are narrowly triangular, 60–80 mm long, soft in texture and covered with soft hairs, and jointed at the base.

DESCRIPTION, REPRODUCTIVE PARTS

Male cones of *Cycas clivicola* subsp. *lutea* are solitary, narrowly ovoid in shape, yellow to brown in colour, 20–25 cm long and 8–9 cm in diameter. Microsporophylls have a firm blade that is not dorsiventrally thickened, 3–35 mm long, 18–19 mm wide, with a fertile zone 25–30 mm long and a sterile apex that is level (not raised), 5–6 mm long, and with a prominent, sharply upturned apical spine 5–6 mm long. Megasporophylls are 12–18 cm long, grey, yellow or pink-tomentose with 2–4 glabrous ovules; the megasporophyll blade is disc-shaped, 90–120 mm long, 45–70 mm wide, moderately pectinate with 24–40 soft lateral spines 25–30 mm long, 1.5–2 mm wide, with an apical spine distinct from the lateral spines, 45–75 mm long by 3–5 mm wide at the base. Seeds are flattened-ovoid, about 38 mm long by 28 mm wide, having an orange sarcotesta, 3 mm thick, not powdery and with a fibrous layer, and a smooth sclerotesta with no internal spongy layer.

DISTINGUISHING FEATURES

Cycas clivicola subsp. *lutea* is distinguished from other Vietnamese cycads by its smooth, yellowish trunks, and its relative small male cones that have small, short microsporophylls and thickened axes. The associated subspecies *clivicola* has even smaller male cones with smaller microsporophylls.

CONSERVATION STATUS

Cycas clivicola subsp. *lutea* is abundant and widespread, and is located in generally inaccessible sites on steep limestone cliffs.

The population number and area of occurrence for the subspecies are not known; the current IUCN Red List status assignment for the two subspecies collectively is NT (near threatened) (Hill et al. 2003).

HYBRIDS AND INTERGRADES

None reported.

USEFUL LITERATURE

Hill et al. (2004) pp. 173–174

Jones (2002) p. 157

Tang et al. (1997) p. 15

Whitelock (2002) pp. 102–103.

FIGURE 40 (a). ***Cycas clivicola*** **subsp.** ***lutea*****. Older plants on vertical limestone cliffs have procumbent or J-shaped trunks, as with this example, near Aranyaprathet, Prachinburi Province, Thailand. Photo: Ken Hill.**

FIGURE 40 (b). Male cone, plant in cultivation, Nong Nooch Tropical Garden, Thailand. Photo: Anders Lindström.

FIGURE 40 (c). Cluster of megasporophylls showing cycad aulacaspis scale (CAS) infestation, plant on Thai-Cambodia border. Photo: Willie Tang.

FIGURE 40 (d). Leaf detail, plant 98975P, in cultivation, Montgomery Botanical Center, Florida, USA. Photo: Jody Haynes.

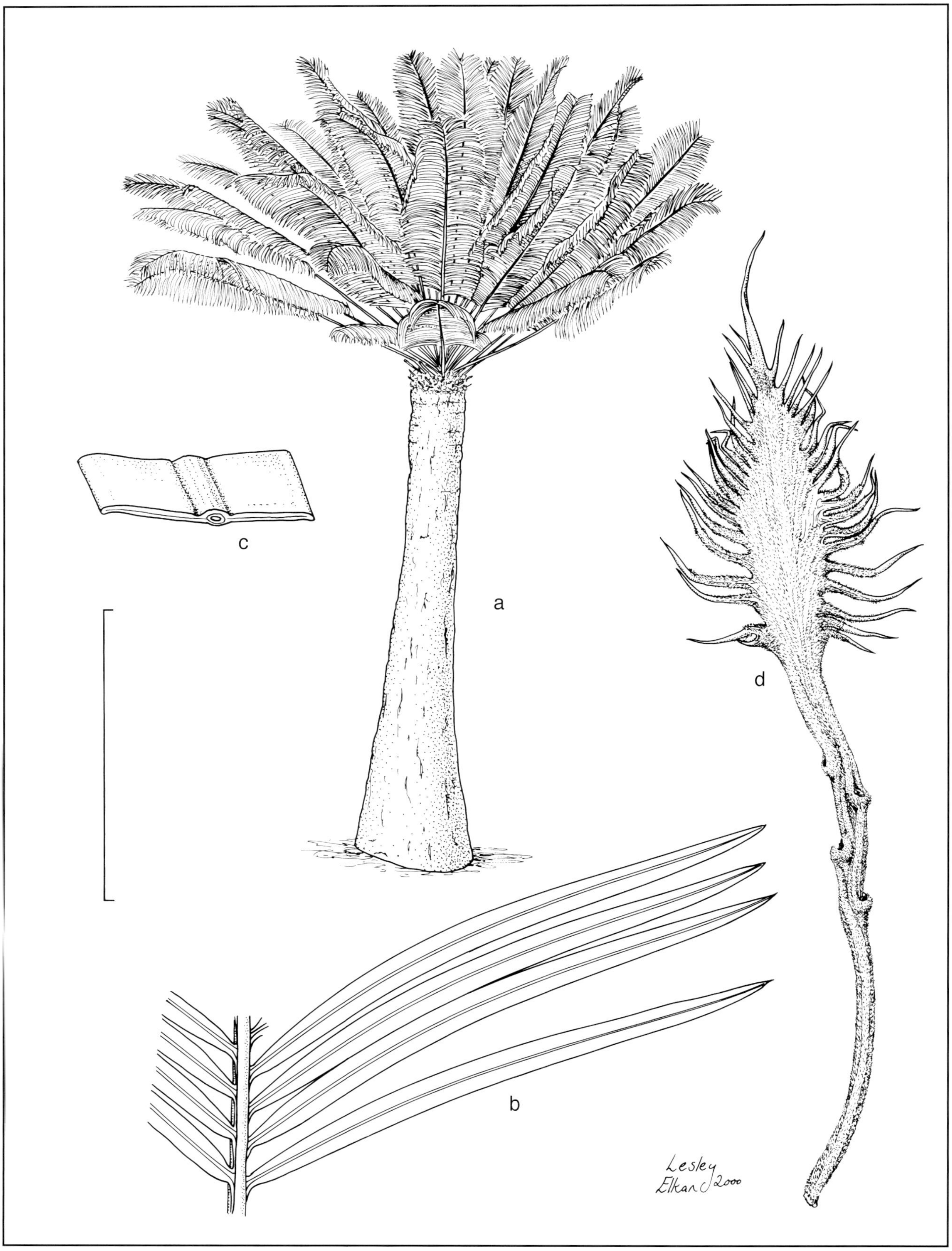

FIGURE 41 (a) *Cycas elongata.* Line drawing by Lesley Elkan. Bar a = 80 cm, b, d = 8 cm, c = 1 cm. Reproduced from Hill et al. (2004) by permission.

21. *Cycas elongata* (Leandri) D. Yue Wang

Publication

D. Y. Wang. 1996. In: F. X. Wang & H. B. Laing (eds), Cycads in China, pp. 51–54, plate 6, fig. 1, as *Cycas elonga*.

Synonyms

Cycas pectinata var. *elongata* (Leandri 1931) and *Epicycas elongata* (Leandri) S. L.Yang (in part, de Laubenfels & Adema 1998). The name *C. inermis* Lour. has been misapplied to this species (Ho 1991).

Etymology

From the Latin *elongatus*, apparently in reference to the elongated apical spine on the megasporophylls of this species (*but see discussion on page 85).

Discovery

The French botanist Jacques Leandri recognised that a population of cycads in southern Vietnam had similarities to the widespread *Cycas pectinata* and published the taxon *C. pectinata* var. *elongata* in 1931. Leandri cited specimens in the Paris Museum of Natural History in his description; these had been collected by Poilane near Ca Na in Ninh Thuan Province. In the 1990's, Si Lin Yang studied these plants in the wild while Ding Yue Wang assessed the ex situ specimens at the Shenzhen Fairy Lake Botanic Garden in China. Wang (1996) then elevated the taxon to specific level as '*C. elonga*', corrected in errata to *C. elongata*. The publication by Hill et al. (2004) validates the name *C. elongata* (Leandri) D.Yue Wang, designating the Paris herbarium (P) voucher *Poilane 9328* as the lectotype.

A proposal to segregate the species into the genus *Epicycas* de Laub. (de Laubenfels & Adema 1998) has been rejected (Chen et al. 2004).

Distribution and habitat

Cycas elongata is locally abundant southern Vietnam's coastal hills, occurring in a narrow discontinuous band about 250 km long extending from the northernmost part of Ninh Thuan Province, west of Cam Ranh in Khanh Hoa Province, in the Song Cau district of Phu Yen Province to near the Cu Mong Pass at the southern border of Binh Dinh Province.
It grows on east-facing slopes at altitudes from 50 to 200 m, on gritty soils derived from coarse siliceous granite. The sites are subjected to periodic fires. The habitat of this species was originally forest and open shrubland, but much has now been degraded to mixed thorny scrub.

Description, vegetative parts

Stems of *Cycas elongata* are aerial, unbranched or occasionally branched, 2–5 m tall, 10–20 cm in diameter and bear 30–60 leaves in the crown. In older specimens the trunks take on a smooth appearance and become progressively thicker at the base, diameters increasing to 1 m at ground level, but without being bulbous. Leaves are bright green or grey-green, semi-glossy, with a white tomentum that is lost as the leaf expands. They are 90–140 cm long, moderately keeled in section (opposing leaflets at 90–150° across the rachis). Leaves have 130–240 leaflets and the rachis ends with a spine or leaflet pair. The petiole is 20–40 cm long (20–30% of total leaf length), glabrous and spinescent for 60–100% of its length. Basal leaflets are 45–60 mm long and are abruptly replaced by petiolar spines. Median leaflets are simple, strongly discolorous, 140–190 mm long, 8–11 mm wide, angled forward (inserted at 50–60° to the rachis), decurrent for 3–5 mm, narrowed to 2.5–4.5 mm at the base, 10–18 mm apart on the rachis, flat in section with flat margins, with the apex acute but not spinescent, and with the midrib raised above and below. Cataphylls are narrowly triangular, soft with a sparse covering of hairs or lacking any tomentum, and 60–90 mm long.

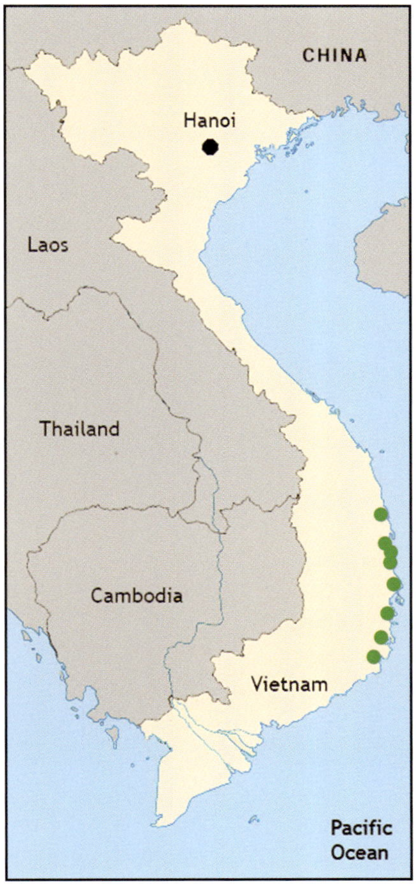

Cycas elongata **– distribution**

FIGURE 41 (b). Coastal hillside habitat for *Cycas elongata*, now degraded to mixed thorny scrub, near Cam Rahn, Khanh Hoa Province. Photo: Ken Hill.

FIGURE 41 (c). Male cone, plant in cultivation, Ba Ngoi town, Khanh Hoa Province. Photo: Roy Osborne.

FIGURE 41 (d). Plant in partially degraded habitat, Song Cau district, Phu Yen Province. Photo: Roy Osborne.

FIGURE 41 (e). Leaf detail, plant near Cam Lap village, Khanh Hoa Province. Photo: Roy Osborne.

FIGURE 41 (f). Megasporophyll cluster on female plant in Ba Ngoi town. Photo: Roy Osborne.

DESCRIPTION, REPRODUCTIVE PARTS

Male cones of *Cycas elongata* are solitary, narrowly ovoid in shape, orange or brown in colour, 25–30 cm long and 9–13 cm in diameter. Microsporophylls have a firm blade that is not dorsiventrally thickened. Megasporophylls are 20–27 cm long, grey-tomentose or brown tomentose with 2–6 glabrous ovules; the megasporophyll blade is ovate, 130–180 mm long, 50–70 mm wide, moderately to deeply pectinate with 36–38 soft lateral spines 20–25 mm long, 1–2 mm wide with an apical spine distinct from the lateral spines, 30–55 mm long by 4–5 mm wide at the base. Seeds are flattened-ovoid, about 40mm long by 30 mm wide, having a yellow sarcotesta, with a fibrous layer, and a smooth sclerotesta with no internal spongy layer.

DISTINGUISHING FEATURES

Cycas elongata is nearest to *C. pectinata*, from which it is distinguished by its keeled leaves with broader leaflets and by its narrow megasporophyll apex with shorter lateral spines. Male cones of *C. elongata* are smaller than those of *C. pectinata* and have shorter microsporophylls with shorter apical spines. Its relatively smooth trunk surface separates *C. elongata* from *C. pachypoda* while the absence of a prominent bulbous base distinguishes *C. elongat*a from both *C. pachypod*a and *C. siamensis*.

[Use of the epithet '*elongata*', in reference to an elongated megasporophyll spine for *Cycas elongata*, is the cause of some confusion. Fieldwork has now shown that both the apical and lateral megasporophyll spines of *C. elongata* are typically shorter than those on *C. pectinata*. Osborne & Nguyen (2002) speculate that the apical-to-lateral spine length proportion may be a more useful character; this is usually about 2:1 in *C. elongata* and about 1:1 in *C. pectinata*.]

CONSERVATION STATUS

Populations of *Cycas elongata* have been reduced significantly through large-scale clearing for agriculture and through removal of specimen plants for trade. An additional threat is the occasional decapitation of female megasporophyll clusters for use in decoration (Osborne & Nguyen 2002). The currently estimated population numbers exceed 10 000, the area of occurrence is about 100 km^2 and the current IUCN Red List status assignment is VU (vulnerable) (Hill et al. 2003).

HYBRIDS AND INTERGRADES

The northernmost populations of *C. elongata* may intergrade with *C. pectinata* in the southern area of Binh Dinh Province.

USEFUL LITERATURE

de Laubenfels & Adema (1998, p. 393, as *Epicycas elongata*)
Hill et al. (2004) pp. 174–176
Jones (2002) pp. 166–167
Osborne & Nguyen (2002)
Wang (1996) pp. 51–53
Whitelock (2002) pp. 110–111.

FIGURE 41 (g). Wild-collected *Cycas elongata* plants being trans-ported from Ba Ngoi for sale in Nha Trang City. Photo: Roy Osborne.

FIGURE 41 (h). Wild-collected plants used in a mass planting in front of a popular tourist hotel in Nha Trang. Photo: Piet Vorster.

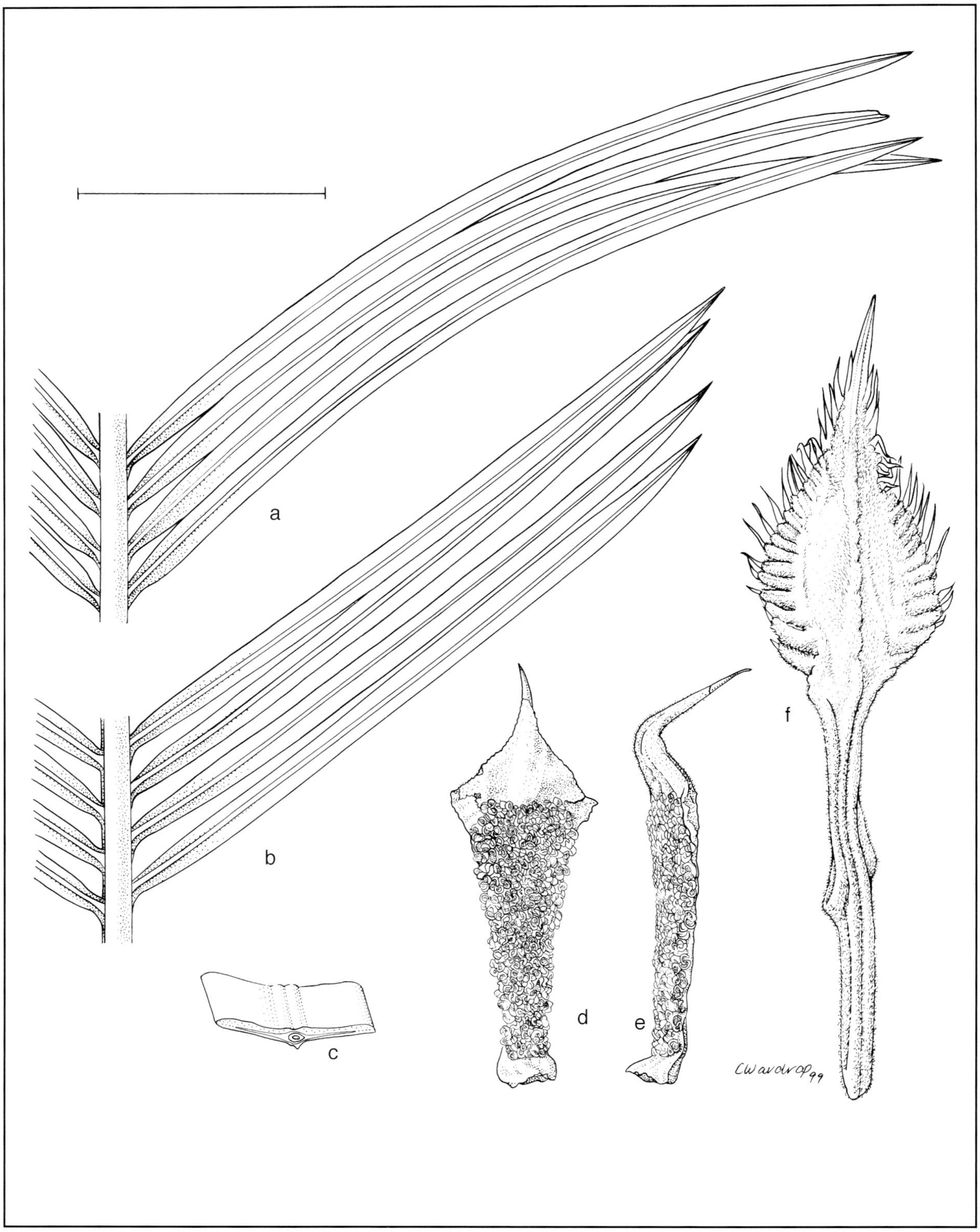

FIGURE 42 (a) *Cycas pachypoda*. Line drawing by Catherine Wardrop. Bar a, b, d = 5 cm, e, f = 2.5 cm, c = 1 cm. Reproduced from the species description by permission.

22. *Cycas pachypoda* K. D. Hill

PUBLICATION

K. D. Hill, Hiêp T. Nguyen & Phan K. Loc. 2004. The Botanical Review 70(2): 176–178, fig. 13.

SYNONYMS

None.

ETYMOLOGY

From the Greek παχυς (*pachys*), thick, and ποδιον (*podion*), foot, in reference to the distinctive large swollen trunk bases of mature plants of this species.

DISCOVERY

Like many other cycads in this volume, *Cycas pachypoda* was recognised as a distinct species only after fieldwork in the 1990s. The Chinese scientist Si Lin Yang obtained material from at Ca Na, in Ninh Thuan Province, in 1994 while Ken Hill collected the holotype (*Hill 5012*) on 20 May 1998 from Vinh Hao village, Tuy Phong district in Binh Thuan Province. This is filed at the herbarium of the National Center for Natural Sciences and Technology, Hanoi (HN) with isotypes deposited at herbaria in Bangkok (BKF), Beijing (PE), Kew (K), Leiden (L), Missouri (MO), New York (NY), Paris (P) and Sydney (NSW). The species description was published in 2004.

The Vietnamese name for *C. pachypoda* is thiên tuế chân voi– chân voi (elephant leg cycad).

DISTRIBUTION AND HABITAT

Cycas pachypoda occurs in southern Vietnam's coastal hills, populations being locally abundant but having a restricted distribution in Binh Thuan and Ninh Thuan Provinces*. It grows on rocky south-facing slopes on dry, gritty soils derived from coarse siliceous granite. The habitat from this species was originally low monsoon woodland but this has now largely been degraded to open thorn shrubland as a result of agricultural and fuel-collecting activities. Plant populations are subject to periodic fires.

[Hill et al. (2004) include Dong Nai Province in the area of occurrence but there is no real evidence that the species is found that far inland.]

DESCRIPTION, VEGETATIVE PARTS

Stems of *Cycas pachypoda* are aerial, 0.5–2 m tall, 12–17 cm in diameter above but abruptly swollen towards ground level, with older specimens having thick, fissured, corky bark, and bear 50–80 leaves in the crown. Leaves are deep green or grey-green, semi-glossy, with a white tomentum that is lost as the leaf expands. They are 90–130 cm long, slightly keeled or flat in section (opposing leaflets at 150–170° across the rachis). Leaves have 160–250 leaflets and the rachis usually end with a spine 5–15 mm long. The petiole is 10–20 cm long (10–20% of total leaf length), glabrous and spinescent for 80–100% of its length. Basal leaflets are short (15–45 mm long) but not gradually reducing to petiolar spines. Median leaflets are simple, strongly discolorous, 120–180 mm long, 6–7 mm wide, angled forward (inserted at 50–60° to the rachis), decurrent for 3–5 mm, narrowed to 2.5–3.5 mm at the base, 6–12 mm apart on the rachis, flat or slightly keeled in section with flat or slightly recurved margins, with the apex acute and spinescent, and with the midrib raised above and below. Cataphylls are narrowly triangular, soft with a sparse covering of hairs or lacking any tomentum, and 80–110 mm long.

DESCRIPTION, REPRODUCTIVE PARTS

Male cones of *Cycas elongata* are solitary, narrowly ovoid in shape, orange or brown in colour, 25–40 cm long and 10–15 cm in diameter. Microsporophylls have a firm blade that is not dorsiventrally thickened, 40–52 mm long, 17–20 mm wide with a fertile zone 30–40 mm long and a sterile apex 7–12 mm long, level but with

Cycas pachypoda **– distribution**

FIGURE 42 (b). *Cycas pachpoda* on rocky terrain at the foot of coastal hills, near Vinh Hoa village, Binh Thuan Province. Photo: Ken Hill.

FIGURE 42 (c). Plant near Ca Na village, Ninh Thuan Province. Photo: Roy Osborne.

FIGURE 42 (d). Leaf detail, plant near Ca Na Village. Photo: Piet Vorster.

FIGURE 42 (e). Cluster of megasporophylls, plant near Vinh Hoa village. Note prominent and distinct apical spine on each megasporophyll. Photo: Ken Hill.

FIGURE 42 (f). Male cone on a plant near Ca Na village. Photo: Roy Osborne.

Left – FIGURE 43 (g). Wild collected plants being offered for sale on the highway near Ca Na village, Ninh Thuan Province. Photo: Roy Osborne.

a sharply upturned apical spine 9–15 mm long. Megasporophylls are 7–15 cm long, brown-tomentose, with 2–4 glabrous ovules; the megasporophyll blade is disc-shaped, 90–120 mm long, 40–70 mm wide, deeply pectinate with 30–36 soft to sharp lateral spines 10–20 mm long, 1–2 mm wide, with an apical spine distinct from the lateral spines, 15–40 mm long by 4–5 mm wide at the base. Seeds are flattened-ovoid, having a yellow sarcotesta, with a fibrous layer, and a smooth sclerotesta with no internal spongy layer.

DISTINGUISHING FEATURES

Cycas pachypoda is characterised by trunks with a very broad and abruptly swollen base, with thick, fissured, corky bark, and by its slightly keeled to flat leaves with short petioles and stiff, narrow leaflets, its medium sized male cones with short apical spines on the microsporophylls, and its generally small megasporophylls. *C. siamensis* is similar in having a swollen base and thick bark, but has smaller male cones, flat leaves with a more persistent tomentum and has lower leaflets gradually reducing to spines. *C. pachypoda* is distinguished from *C. clivicola* and *C. condaoensis* by having short and mostly spinescent petioles, and larger male cones with longer microsporophylls.

CONSERVATION STATUS

Populations of *Cycas pachypoda* have been reduced significantly through habitat degradation, large-scale clearing for agriculture and through removal of specimen plants for trade. The current estimated total population number falls in the 2 500–10 000 range although the area of occurrence is as little as 20 km^2; the present IUCN Red List status assignment is VU (vulnerable) (Hill et al. 2003).

HYBRIDS AND INTERGRADES

None known. The *Cycas pachypoda* population at Ca Na has plants occurring within meters of *C. lindstromii* specimens. Despite the fact that both species cone intermittently throughout the year, there is no evidence of any hybridisation between them; the barrier may be due to species-specific pollinators or may be genetic (Tang 2003).

USEFUL LITERATURE

Hill et al. (2004) pp. 176–178
Jones (2002) p. 191
Whitelock (2002) p. 132.

23. *Cycas condaoensis* K. D. Hill & S. L. Yang

PUBLICATION

K. D. Hill, Hiêp T. Nguyen & Phan K. Loc. 2004. The Botanical Review 70(2): 178–179, fig. 14.

SYNONYMS

Early reports cited *Cycas pectinata* Buch.-Ham. (Leandri 1931) and *C. inermis* Lour. (Hiêp & Vidal 1996) in reference to this cycad.

ETYMOLOGY

Named in reference to the occurrence of this species on the Con Dao archipelago, off the south-east coast of Vietnam.

DISCOVERY

Although *Cycas condaoensis* was named only in 2004, the endemic cycad of the Con Dao islands (previously Poulo Condore) has been known for many years. The nominated holotype, Kew voucher *L. Pierre 1772*, was collected as early as September 1877, from Con Son, the largest of the islands (infamous as an early French prison island); isotypes are present at the Arnold Arboretum of Harvard University, Massachusetts, USA (A), at Paris (P), Leiden (L) and New York (NY). A specimen (*Averyanov and Kudryavtzeva 361)* is filed at the V. L. Komarov Botanical Institute in Saint Petersburg, Russia (LE). In addition, the Belgian plant collector, Louis van Houtte, arranged for several live plant specimens to be sent to Kew in the 1880s.

C. condaoensis is known locally as thiên tuế Côn Đảo (Con Dao cycad).

DISTRIBUTION AND HABITAT

Cycas condoaensis is endemic in the Con Dao islands, which lie approximately 100 km offshore from the south-eastern part of the Vietnamese mainland. These sixteen islands, of which Con Son (5 150 ha) is the largest, comprise part of Ba Ria-Vung Tau Province. Several discrete *C. condaoensis* populations

FIGURE 43 (a). *Cycas condaoensis.* Line drawing by Lesley Elkan. Bar a = 100 cm, b = 8 cm, c = 6 cm. Reproduced from the species description by permission.

are known, typically occurring on stabilised sand dunes. The habitat ranges from open low shrubland, to dense tall shrubland, or woodland.

DESCRIPTION, VEGETATIVE PARTS

Stems of *Cycas condaoensis* are aerial, 0.2–2.5 m tall, 14–17 cm in diameter above but swollen towards ground level, with older specimens having thick, corky bark, and bearing 12–70 leaves in the crown. Leaves are bright green, semi-glossy, with a white or brown tomentum that is lost as the leaf expands. They are 70–175 cm long, flat (not keeled) in section (opposing leaflets at 180° across the rachis). Leaves have 130–260 leaflets and the rachis usually ends with a leaflet pair. The petiole is 20–50 cm long (15–30% of total leaf length), glabrous and spinescent for 0–60% of its length. Basal leaflets are 25–120 mm long and do not gradually reduce to petiolar spines. Median leaflets are simple, 130–240 mm long, 7–11.5 mm wide, angled forward (inserted at 60–75° to the rachis), decurrent for 3–6 mm, narrowed to 2–3 mm at the base, 5–17 mm apart on the rachis, flat in section with slightly recurved margins, with the apex acute but not spinescent, and with the midrib raised above and below. Cataphylls are narrowly triangular, sharp, with a sparse covering of hairs or lacking any tomentum, 50–60 mm long and long lasting.

DESCRIPTION, REPRODUCTIVE PARTS

Male cones of *Cycas condaoensis* are solitary, narrowly ovoid in shape, orange to brown or green in colour, 26–31 cm long and 9–12 cm in diameter. Microsporophylls have a firm blade that is not dorsiventrally thickened, 30–50 mm long, 12–20 mm wide with a fertile zone 23–36 mm long and a sterile apex 7–11 mm long, level but with a sharply upturned apical spine 4–5 mm long. Megasporophylls are 9–17 cm long, yellow or pale brown-tomentose, with 2–4 glabrous ovules; the megasporophyll blade is disc-shaped, 60–120 mm long, 35–55 mm wide, deeply pectinate with 20–32 soft lateral spines 8–18 mm long, 1.5–2 mm wide, with an apical spine distinct from the lateral spines, 25–60 mm long by 4–10 mm wide at the base. Seeds are ovoid, 39–43 mm long, 33–36 mm wide, have a smooth, not powdery, yellow to orange sarcotesta 4–6 mm thick, with a fibrous layer, and with no internal spongy layer.

DISTINGUISHING FEATURES

Cycas condaoensis shares many characters with *C. siamensis* and the *C. pectinata* group, all these taxa having a swollen trunk at the base and a thick corky bark. Its small, oblong male cone separates *C. condaoensis* from members of the *C. pectinata* group, while the glabrous leaf with basal leaflets not reducing to spines distinguishes it from *C. siamensis*. Male cones of *C. condaoensis* are intermediate in size between those of *C. siamensis* and *C. pachypoda*.

CONSERVATION STATUS

All populations of *Cycas condaoensis* fall within the Con Dao National Park; the populations are stable and habitat reduction is not a threat. The total population number of this species is estimated at 2 500–10 000. However, the restricted area of occurrence (only about 20 km^2) and the limited distribution of this species warrants its listing as VU (vulnerable) in the current IUCN Red List (Hill et al. 2003).

HYBRIDS AND INTERGRADES

None known. Being isolated by about 100 km from the mainland, there is little opportunity for any genetic exchange with any other cycads.

USEFUL LITERATURE

Hill et al. (2004) pp. 178–179
Jones (2002) pp. 157–158
Whitelock (2002) pp. 103–104.

Cycas condaoensis **– distribution**

FIGURE 43 (b). Stabilised sand dune habitat for *Cycas condaoensis*, Con Son Island, Con Dao Archipelago. Photo: Ken Hill.

FIGURE 43 (c). *Cycas condaoensis* in habitat, Con Son Island. Photo: Ken Hill.

Left – FIGURE 43 (d). Leaf detail, plant in cultivation at Nong Nooch Tropical Garden, Thailand. Photo: Anders Lindström.

FIGURE 43 (e). Male cone, Con Son Island. Note the sharply upturned apical spine on each microsporophyll. Photo: Ken Hill.

Left – FIGURE 43 (f). Cluster of megasporophylls shortly after fertilisation. Note the pubescent lamina and the distinct apical spine. Con Son Island. Photo: Ken Hill.

24. *Cycas lindstromii* S. L. Yang, K. D. Hill & N. T. Hiêp

PUBLICATION

S. L. Yang, K. D. Hill & N. T. Hiêp. 1997. Novon 7: 213–215, fig. 1.

SYNONYMS

The name *Epicycas lindstromii* (S. L.Yang, K. D. Hill & N. T. Hiêp) de Laub. has been used for this species (de Laubenfels & Adema 1998).

ETYMOLOGY

Named for the Swedish horticulturist Anders Lindström, collections manager of the Nong Nooch Tropical Botanical Garden, Chonburi, Thailand, who has contributed greatly to the knowledge of Asian and other cycads and who participated in the discovery of this species.

DISCOVERY

During fieldwork in southern Vietnam in 1994, Lindström and Si Lin Yang noticed vendors selling unusual plants alongside Highway 1 in Binh Thuan Province; subsequent exploration revealed nearby populations of a new dwarf cycad with a subterranean caudex. The species was described in 1997 by Yang, Hill and Hiêp as *Cycas lindstromii*. The holotype, *S. L. Yang 557*, collected in November 1994 near Doc Bao Da village, Xa Chi Cong, Tuy Phong district in Binh Thuan Province, is held at Fairchild Tropical Garden, Florida, USA (FTG) with an isotype at the Royal Botanic Gardens, Sydney, Australia (NSW). The species also corresponds to a herbarium specimen collected near Nha Trang as early as 1911 (Robinson 1304), filed at New York and later at Kew (K).

A proposal to segregate the species into the genus *Epicycas* de Laub. (de Laubenfels & Adema 1998) has been rejected (Chen et al. 2004).

DISTRIBUTION AND HABITAT

Cycas lindstromii is found in the hottest and driest parts of Vietnam, in low, flat sandy sites along the south-eastern coast. Scattered populations have been recorded from Ba Ria-Vung Tau, Binh Thuan, Khanh Hoa and Ninh Thuan Provinces. The cycads grow in grasslands and mixed, dry, open forest or woodland often dominated by *Dipterocarpus tuberculatus*. The soil is deep and sandy, arising from siliceous granite outwash and beach dune sands.

DESCRIPTION, VEGETATIVE PARTS

Stems of *Cycas lindstromii* are subterranean, typically 5–8 cm in diameter but markedly swollen at the base and often branched towards the apex, with a smooth surface, white-grey in colour, and bearing 6–25 leaves in the crown. Leaves are bright green above but yellowish green beneath, highly glossy, with a brown tomentum that is lost as the leaf expands. They are 40–100 cm long, moderately keeled in section (opposing leaflets at 90–130° across the rachis). Leaves have 50–140 leaflets and the rachis consistently ends with a leaflet pair. The petiole is 15–35 cm long (25–40% of total leaf length), glabrous and spinescent for 50–100% of its length. Basal leaflets are 35–120 mm long) and do not gradually reduce to petiolar spines. Median leaflets are simple, strongly discolorous, 100–170 mm long, 7–11 mm wide, angled forward (inserted at 45–60° to the rachis), decurrent for 3–5 mm, narrowed to 2–3 mm at the base, 7–13 mm apart on the rachis, flat in section with flat margins, with the apex acute but not spinescent, and with the midrib raised above and below. Cataphylls are linear, 20–50 mm long, 3–7 mm wide at the base, soft in texture with a covering of soft hairs, and long-lasting.

DESCRIPTION, REPRODUCTIVE PARTS

Male cones of *Cycas lindstromii* are solitary, spindle-shaped, yellow in colour, 12–20 cm long and 4–5 cm in diameter. Microsporophylls have

Cycas lindstromii **– distribution**

FIGURE 44 (a). Coastal hills near Ca Na village, Ninh Thuan Province. This habitat is shared by *C. lindstromii* (on the lower slopes and flatter areas) and *C. pachypoda* (on the hillside). Photo: Roy Osborne.

FIGURE 44 (b). *Cycas lindstromii* on coastal flats near Ca Na village. Photo: Roy Osborne.

FIGURE 44 (c). A vigorous new leaf flush on a plant near Ca Na. Photo: Jeff Chemnick.

FIGURE 44 (d). Leaf detail, plant near Ca Na village. Photo: Piet Vorster.

FIGURE 44 (e). Male cone. *Cycas lindstromii* near Vinh Hao village, Binh Thuan Province. Photo: Ken Hill.

FIGURE 44 (f). Wild collected *Cycas lindstromii* from the Ca Na area being offered for sale in Hué City. Photo: Roy Osborne.

a firm blade that is not dorsiventrally thickened, 10–16 mm long, 8–15 mm wide with a fertile zone 7–12 mm long and a sterile apex 3–4 mm long, level, with a rudimentary sharply upturned apical spine 2–8 mm long. Megasporophylls are 8–20 cm long, grey-tomentose, with 2–4 glabrous ovules; the megasporophyll blade is ovate, 50–100 mm long, 15–40 mm wide, deeply pectinate with 8–26 soft lateral spines 8–30 mm long, 1.5–3 mm wide, with an apical spine distinct from the lateral spines, 10–50 mm long by 1.5–4 mm wide at the base. Seeds are ovoid, 29–37 mm long, 20–33 mm wide, have a smooth, not powdery, orange sarcotesta 3.5–4 mm thick, with a fibrous layer, and with no internal spongy layer.

FIGURE 44 (g). *Cycas lindstromii* megasporophylls with seeds. Plant near Vinh Hao village. Photo: Ken Hill.

DISTINGUISHING FEATURES

Cycas lindstromii is characterised by its subterranean and much-branched trunk, its short, keeled leaves with basal leaflets not being reduced to spines, its small, slender male cones, and its relatively long and narrow megasporophyll blade. Despite its apparent gross differences to other members of *C. pectinata* group, this species is considered to be part of that group on account of its pectinate megasporophyll and its fibrous sarcotesta. *C. lindstromii* appears to be most closely related to *C. pranburiensis*, a dwarf coastal species from peninsular Thailand.

CONSERVATION STATUS

Extensive plant removal for horticulture and considerable habitat degradation both pose threats for the future of *Cycas lindstromii*. One population has the protection of being within the Binh Chau-Phuoc Buu Nature Reserve in Ba Ria-Vung Tau Province. The present total population number for the species exceeds 10 000 individuals in an area of occurrence of about 200 km^2 and the current IUCN Red List status is VU (vulnerable) (Hill et al. 2003).

HYBRIDS AND INTERGRADES

None known. The *Cycas lindstromii* population at Ca Na has plants occurring within meters of *C. pachypoda* specimens. Despite the fact that both species cone intermittently throughout the year, there is no evidence of any hybridisation between them; the barrier may be due to species-specific pollinators or may be genetic (Tang 2003).

USEFUL LITERATURE

de Laubenfels & Adema (1998, p. 395, as *Epicycas lindstromii*)
Hill et al. (2004) pp. 179–180
Jones (2002) pp. 174–175
Tang (2003)
Whitelock (2002) pp. 119–120
Yang et al. (1997).

25. *Cycas siamensis* Miq.

PUBLICATION

F. A. W. Miquel. 1863. Botanische Zeitung (Berlin) 21: 334.

SYNONYMS

Synonyms for *Cycas siamensis* include *Epicycas siamensis* (Miq.) de Laub. (de Laubenfels & Adema 1998), *Cycas cochinchinensis* Warb. ex Schuster (1932), *C. intermedia* Hort. ex B. S. Williams (as cited in Hiêp & Loc 1999) and *C. immersa* Craib (1912). The name has also been widely misapplied to many short, squat cycads from south-east Asia, including several Chinese taxa.

ETYMOLOGY

The epithet for *Cycas siamensis* refers to its first collection in Siam (now Thailand).

DISCOVERY

The Dutch plant collector Johannes Teijsmann came across an unknown cycad in March 1862 during travels through what is now the Kanchanaburi Province of Thailand. He sent specimens to botanist Friederich Miquel, who described the species in 1863 citing the holotype as two unnumbered herbarium sheets (*J. E. Teijsmann*)

FIGURE 45 (a). *Cycas siamensis* in habitat, Uttaradit Province, Thailand, with Willie Tang. Photo: Si Lin Yang.

FIGURE 45 (b). Plants in habitat, Tak Province, Thailand, with Si Lin Yang. Photo: Willie Tang.

FIGURE 45 (c). Plant with male cone, central highlands, Dac Lak Province. Photo: Si Lin Yang.

FIGURE 45 (d). Leaf detail, from a plant in Tak Province. Thailand. Photo: Willie Tang.

FIGURE 45 (e). Male cone, plant near Tha Takiap village, Chacgoengsao Province, Thailand. Photo: Ken Hill.

deposited in the Utrecht herbarium (U). The range become progressive more extended as reports appeared of the same species in many other provinces in the northern half of Thailand; further collections showed that *Cycas siamensis* also occurs in Cambodia and Myanmar. In the 1990s, Si Lin Yang and Hiêp Nguyen found disjunct populations in central and northern Vietnam.

A proposal to segregate the species into the genus *Epicycas* de Laub. (de Laubenfels & Adema 1998) has been rejected (Chen et al. 2004).

DISTRIBUTION AND HABITAT

Cycas siamensis is distributed widely through Thailand and also occurs in Cambodia, Myanmar and Vietnam. The distribution almost certainly extends to Laos but probably not to China. There are two principal Vietnamese localities: in the in the central Annam Highlands (Dac Lak and Kon Tum Provinces), and less commonly in the north (Nghe An Province)*. Plants are found often in dense stands, in full sun to light shade, in low open woodland on flat country or low hills, usually at altitudes below 300 m. Soils are usually shallow and stony and may be limestone-derived or lateritic. Associated trees are *Dipterocarpus tuberculatus*, *Erythrina* and *Bombax*. The localities are subject to wet monsoon cycles interspersed with long, dry periods when many of the trees and shrubs, including the cycads, are deciduous, and when fires are common.

Cycas siamensis **– distribution in Vietnam**

[Hiêp and Loc (1999) give a much broader distribution, based on an earlier concept of the species that has since been revised.]

DESCRIPTION, VEGETATIVE PARTS

Stems of *Cycas siamensis* are subterranean to aerial, up to 1.5 m tall and 14–20 cm in diameter, occasionally branched towards the apex but markedly swollen at the base that often becomes broad and plate-like in older specimens. Stems are covered by a thick layer of leaf bases interspersed with dense layers of hair, an adaptive insulation against the fires to which the species is often exposed during the dry seasons (Tang & Yang 1995). Leaves are bright green to deep green, semi-glossy, with a persistent white to orange tomentum. They are 60–120 cm long, flat in section (opposing leaflets at 170–180° across the rachis). Leaves have 140–280 leaflets and the rachis frequently ends with a spine 1–18 mm long. The petiole is 11–13 cm long (15–30% of total leaf length), covered with fine hairs and spinescent for 10–90% of its length. Basal leaflets are 10–50 mm long) and gradually reduce to petiolar spines. Median leaflets are simple, strongly discolorous, 80–145 mm long, 5–8 mm wide, angled forward (inserted at 55–80° to the rachis), decurrent for 2–4 mm, narrowed to 3–5 mm at the base, 5–10 mm apart on the rachis, flat in section with slightly recurved margins, with the apex acutely spinescent, and with the midrib flat to raised above and raised below. Cataphylls are linear, sharp, 60–70 mm long, covered with soft hairs and long-lasting.

DESCRIPTION, REPRODUCTIVE PARTS

Male cones of *Cycas siamensis* are solitary, narrowly ovoid in shape, orange to brown in colour, 10–24 cm long and 5–7.5 cm in diameter. Microsporophylls have a firm blade that is not dorsiventrally thickened, 19–30 mm long, 11–17 mm wide with a fertile zone 13–22 mm long and a sterile apex 5–8 mm long, level, with a prominent, sharply upturned apical spine 7–14 mm long. Megasporophylls are 6–11 cm long, brown-tomentose, with 2 glabrous ovules; the megasporophyll blade is disc-shaped, 60–110 mm long, 30–80 mm wide, deeply pectinate with 22–46 soft lateral spines 8–35 mm long, 1–2 mm wide, with an apical spine distinct from the lateral spines, 23–45 mm long by 4–8 mm wide at the base. Seeds are almost spherical, 30–37 mm long,

26–37 mm wide, have a smooth, not powdery, yellow sarcotesta 2–3 mm thick, with or without a fibrous layer, and with no internal spongy layer.

DISTINGUISHING FEATURES

Cycas siamensis is the last of the Vietnamese section *Indosinenses* cycads dealt with in this volume. Like the other members of *C. pectinata* group, this species shares the characters of a pectinate megasporophyll and a fibrous sarcotesta. Its short-stemmed habit is characteristic, plants coning before any aerial stem is developed; the stem is also recognised by its sharply demarcated bulbous base with persistent leaf bases interspersed with layers of hair. *C. siamensis* typically has flat leaves with a very short petiole, and with basal leaflets which reduce to spines. Male cones have a distinct oblong shape and are smaller that those of most of the *C. pectinata* group. Megasporophylls are distinguished by being usually densely brown-tomentose with the tomentum persisting at the base.

CONSERVATION STATUS

Cycas siamensis extends over an enormous area of south-east Asia, often occurring in dense stands. However, in all areas of its occurrence, the populations are subject to habitat degradation, inappropriate fire regimes, clearance for agriculture and removal of specimen plants for trade. Some Vietnamese populations in Dac Lak Province have the protection of falling within the Yok Don and the Nam Ca Nature Reserves (Hiêp & Loc 1999). Hill et al. (2003) cite a global population number exceeding 10 000 in an area of occurrence of 1000 km^2, but these figures appear conservative. The IUCN Red List status has recently been changed to VU (vulnerable).

HYBRIDS AND INTERGRADES

Tang et al. (1997) mention possible hybridisation between *Cycas siamensis*, *C. pectinata* and *C. nongnoochiae* in Thailand, but there are no reports of any hybrids from the Vietnamese populations.

USEFUL LITERATURE

de Laubenfels & Adema (1998, as *Epicycas siamensis*) pp. 393–394
Hiêp & Loc (1999) p. 29
Hill et al. (2004) pp. 180–182
Jones (2002) pp. 208–209
Miquel (1863)
Norstog & Nicholls (1997) pp. 206–207
Pant (1962)
Smitinand (1972)
Tang & Yang (1995)
Tang et al. (1997)
Whitelock (2002) pp. 146–147
Wang (1996) pp. 114–117.

FIGURE 45 (f). *Cycas siamensis* megasporophyllls with fertile seed, plant near Ban Som Bom, Uttaridit Province, Thailand. Photo: Ken Hill.

26. *Cycas inermis* Lour.

Publication

J. de Loureiro. 1793. Flora Cochinchinensis, second edition, vol. 2: 776–777.

Synonyms

Synonyms for *Cycas inermis* include *C. revoluta* Thunb. var. *inermis* (Lour.) Miq. (Miquel 1861), and *C. siamensis* Miq. subsp. *inermis* (Lour.) J. Schust. (Schuster 1932). The name *C. circinalis* L. was misapplied to this taxon by Hiệp & Vidal (1996) while *C. rumphii* Miq. was similarly misapplied by Leandri (1931) and Ho (1991).

Etymology

From the Latin adjective *inermis*, unarmed or without spines, originally in reference to the absence of petiolar spines for this species. Since this character is not common on representatives of this species, the epithet choice was unfortunate.

Discovery

Cycas inermis was the first of the Vietnamese cycad species to be published in the scientific literature. Jesuit priest and Portuguese naturalist João de Loureiro collected material of a cultivated cycad at an undescribed locality in the area then named Cochin China (now southern Vietnam). In 1793, he wrote a short description of the species for the Flora Cochinchinensis, filing the holotype *Loureiro 3* in the British Natural History Museum (BM). Although this specimen is no more than an apical leaf portion (fig. 6, chapter 2), it closely matches material from forest plants from the Nam Cat Tien National Park, Dong Nai Province. However, Loureiro described the megasporophylls as laciniate (slashed into narrow pointed lobes), which is not consistent with the present concept of *C. inermis*. Hill et al. (2004) speculate that the original description may in fact represent a composite from disparate specimens. During the 1990s, various authors have considered this taxon to be part of *C. circinalis*, *C. revoluta*, *C. rumphii* and *C. siamensis*, but Hill et al.'s (2004) review redefined the species as applying to those inland forest plants from south and central Vietnam that match the type specimen.

Distribution and habitat

Cycas inermis is widely but sporadically distributed on the Vietnamese central and southern near-coastal ranges, with populations in Da Nang, Dong Nai, Khanh Hoa and Quang Nam Provinces. Plants are locally common in closed evergreen or semi-deciduous forest on well-drained slopes or stony rises over granite, metamorphic or basaltic substrates.

Description, vegetative parts

Stems of *Cycas inermis* are aerial, 1.5–4 m tall and 8–14 cm in diameter, and bear 6–35 leaves in the crown. Leaves are deep green, highly glossy, with a white tomentum that is lost as the leaf expands. They are 220–330 cm long, flat (not keeled) in section (opposing leaflets at 180° across the rachis). Leaves have 130–230 leaflets and the rachis consistently ends with leaflet pair. The petiole is 65–80 cm long (25–35% of total leaf length), glabrous and usually spinescent for 70–100% of its length. Wholly unarmed individuals are uncommon. Basal leaflets are abruptly replaced by petiolar spines. Median leaflets are simple, strongly discolorous, 250–350 mm long, 11–18 mm wide, angled forward (inserted at 60–80° to the rachis), decurrent for 5–7 mm, narrowed to 3–4.5 mm at the base, 18–22 mm apart on the rachis, flat in section with flat to slightly recurved margins, with the apex softly tipped (not spinescent), and with the midrib raised above but flush below. Cataphylls are linear, sharp, 40–50 mm long, covered with soft hairs and jointed at the base.

***Cycas inermis* – distribution**

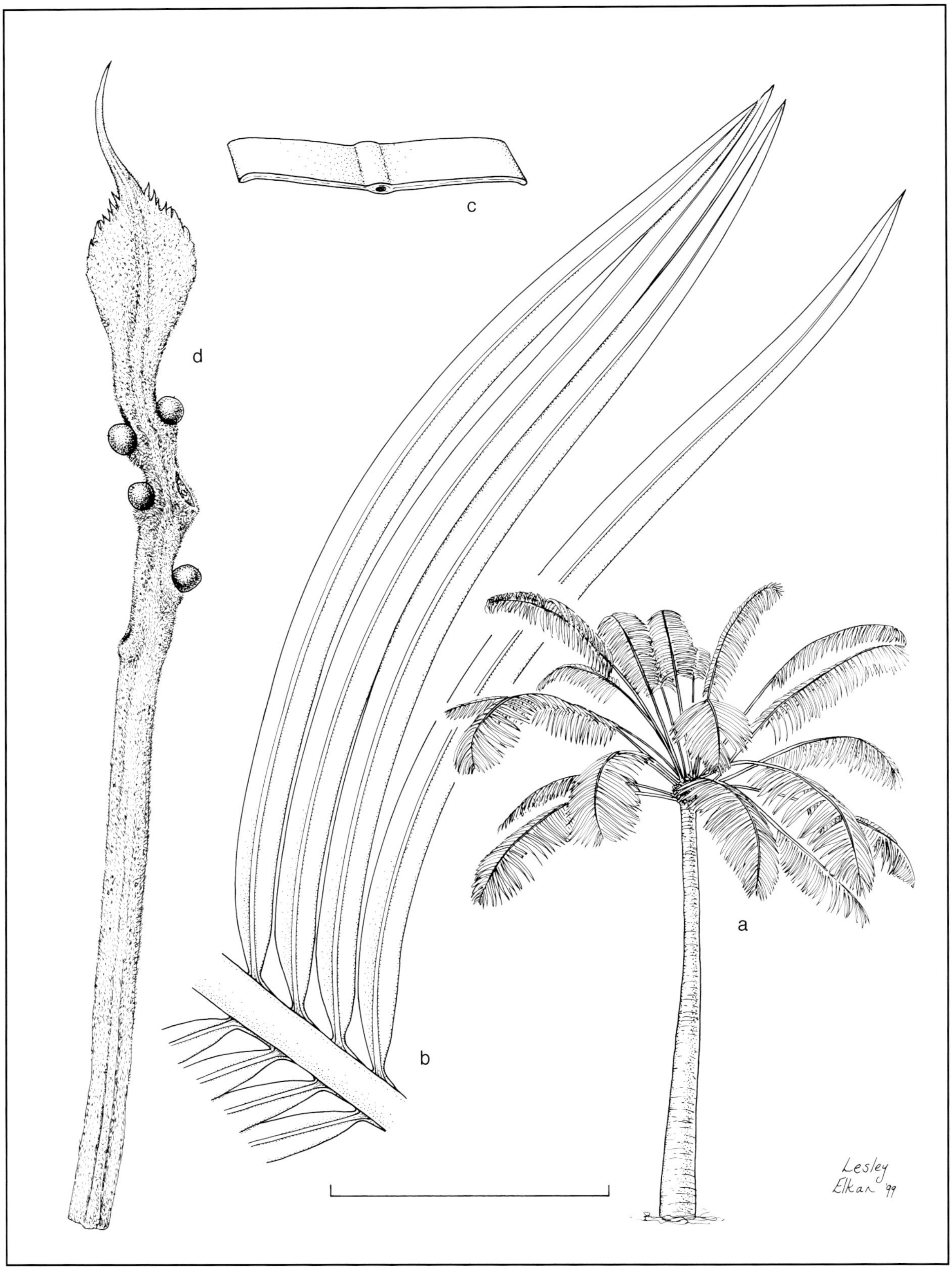

FIGURE 46 (a). *Cycas inermis*. Line drawing by Lesley Elkan. Bar a = 200 cm, b, d = 8 cm, c = 1 cm. Reproduced from Hill et al. (2004) by permission.

DESCRIPTION, REPRODUCTIVE PARTS

Male cones of *Cycas siamensis* are solitary, ovoid in shape, cream in colour, about 12 cm long and 8 cm in diameter. Microsporophylls have a firm blade that is not dorsiventrally thickened, and a prominent, sharply upturned apical spine about 10 mm long. Megasporophylls are 20–30 cm long, grey-tomentose, with 4–8 glabrous ovules; the megasporophyll blade is sword-shaped, 50–100 mm long, 30–65 mm wide, regularly toothed (not pectinate) with 14–26 sharp lateral spines 4–8 mm long, 1–1.5 mm wide, with an apical spine distinct from the lateral spines, 20–30 mm long by 3–4 mm wide at the base. Seeds are oblong, 50–60 mm long, 38–50 mm wide, smooth and slightly powdery, with an orange sarcotesta 4–5 mm thick, with a fibrous layer, but with no internal spongy layer.

DISTINGUISHING FEATURES

Cycas inermis is the only representative of section *Cycas* subsection *Cycas* to occur in Vietnam. The section is defined by its combination of glabrous ovules and non-pectinate megasporophylls while the subsection has a fibrous layer present in the seed sarcotesta. Amongst the Vietnamese cycads, *C. inermis* is readily distinguished by its narrow megaporophyll blade with short lateral teeth, and seeds with a fibrous layer in the sarcotesta. This species appears to be closest to *C. macrocarpa* from Thailand and Malaysia, but differs in having short lateral spines on the megasporophyll.

CONSERVATION STATUS

Cycas inermis extends sporadically over a large north-to-south range. One northern population is protected in the Son Tra Nature Reserve in Da Nang Province, while a southern representative population falls within the borders of the Nam Cat Tien National Park in Dong Nai Province, about 150 km north-east of Ho Chi Minh City. The total population number is estimated to exceed 10 000 in an area of occupancy of about 200 km^2 and the current IUCN Red List status assignment is VU (vulnerable) (Hill et al. 2003).

HYBRIDS AND INTERGRADES

None known.

USEFUL LITERATURE

Hiệp & Loc (1999) p. 29
Hill et al. (2004) pp. 182–186
Jones (2002) pp. 172–173
Whitelock (2002) pp. 116–117.

FIGURE 46 (b). *Cycas inermis* in habitat, low ranges, Cat Tien National Park, Dong Nai Province. Photo: Ken Hill.

FIGURE 46 (c). *Cycas inermis.* Leaf detail, plant at Hai Van Pass, Quang Nam Province. Photo: Ken Hill.

FIGURE 46 (d). *Cycas inermis*. Male cone emerging at stem apex. Photo: Ken Hill.

FIGURE 46 (e). *Cycas inermis*. Seed crop at stem apex. Photo: Ken Hill.

FIGURE 47 (a). *Cycas litoralis* grows amongst low litoral forest on stabilised sand dunes. Type locality, Ao Manao Naol, Thailand. Photo: Ken Hill.

27. *Cycas litoralis* K. D. Hill

PUBLICATION

K. D. Hill & S. L. Yang. 1999b. Brittonia 51: 70–72, fig. 11.

SYNONYMS

Cycas edentata de Laub. is a synonym for *C. litoralis* (in part, de Laubenfels & Adema 1998). The name *C. circinalis* L. was misapplied to the species by Leandri (1931) and Ho (1991), while *C. rumphii* Miq. was similarly misapplied to the taxon by Hiệp & Vidal (1996).

ETYMOLOGY

From the Latin adjective *litoralis*, of the seashore, referring to the preferred beach zone habitat of this cycad.

DISCOVERY

Cycads of section *Cycas* are common feature of shorelines of many tropical and subtropical countries and their associated islands. In 1994, Hill defined subsection *Rumphiae*, with the type *C. rumphii* Miq., on the basis of the spongy layer inside the seeds. This layer provides a flotation device to allow seed dispersal in a marine environment. During fieldwork in south-east Asia in the 1990s, Hill and Yang identified a Thai cycad that was distinct from *C. rumphii* and its allies within subsection *Rumphiae*. The holotype for *C. litoralis* (*Hill 4627*) was collected in April 1994 from Ao Manao Naol, Narathiwat, southern Thailand, and is deposited in the herbarium of the Royal Botanic Gardens, Sydney, Australia (NSW), with isotypes at herbaria in Bangkok (BKF), Kew (K), Leiden (L) and Beijing (PE). The species description was published in 1999.

DISTRIBUTION AND HABITAT

Cycas litoralis occurs along shorelines in southern Myanmar, the south-eastern and south-western coasts of Thailand, peninsular Malaysia and Sumatra. One population is known from Phu Quoc island off the south-western coast of Vietnam in Kien Giang Province. Plants grow in full sun or moderate shade amongst littoral scrub, in beach sand or very shallow soil on rocky granitic or limestone headlands.

DESCRIPTION, VEGETATIVE PARTS

Stems of *Cycas litoralis* are aerial, about 3 m tall but sometimes up to 10 m in height, 11–20 cm in diameter. Leaves are bright green, highly glossy, with a tomentum that is lost as the leaf expands. They are 150–230 cm long, flat (not keeled) in section (opposing leaflets at 180° across the rachis). Leaves have 100–170 leaflets and the rachis consistently ends with leaflet pair. The petiole is 50–90 cm long (30–50% of total leaf length), glabrous and spinescent for 5–60% of its length. Basal leaflets are 170–260 mm long and are abruptly replaced by petiolar spines. Median leaflets are simple, strongly discolorous, 260–350 mm long, 14–19 mm wide, angled forward (inserted at 65–80° to the rachis), decurrent for 8–13 mm, narrowed to 6–8 mm at the base, 15–35 mm apart on the rachis, flat in section with slightly recurved margins, with a sharp apex (not spinescent), and with the midrib flat above but raised below. Cataphylls are narrowly triangular, soft, covered with soft hairs and long-lasting.

DESCRIPTION, REPRODUCTIVE PARTS

Male cones of *Cycas litoralis* are solitary, narrowly ovoid in shape, orange in colour, 35–45 cm long and 13–17 cm in diameter. Microsporophylls have a firm blade that is not dorsiventrally thickened, about 38 mm long by 21 mm wide, with a fertile zone about 29 mm long and a sterile apex about 9 mm long, level, and a prominent, sharply

Cycas litoralis **– distribution close to Vietnam**

FIGURE 47 (b). Well-established *Cycas litoralis* at high-tide mark, Ao Manao Naol, Thailand. Photo: Ken Hill.

FIGURE 47 (c). Leaf detail, plant in cultivation, Plantation 2000, Brisbane, Australia. Photo: Roy Osborne.

upturned apical spine about 24 mm long. Megasporophylls are 30–50 cm long, white- or yellow-tomentose, with 4–8 glabrous ovules; the megasporophyll blade is sword-shaped, 90–120 mm long, 25–40 mm wide, slightly or regularly toothed (not pectinate) with 6–14 soft lateral spines 0–1 mm long, 0–1 mm wide, with an apical spine distinct from the lateral spines, 25–40 mm long by 5–8 mm wide at the base. Seeds are flattened-ovoid, about 52 mm long and 43 mm wide, smooth, not powdery, with an orange-brown sarcotesta 4 mm thick, with no fibrous layer, but a well-developed internal spongy layer.

DISTINGUISHING FEATURES

Cycas litoralis is the only representative of section *Cycas* subsection *Rumphiae* to occur in Vietnam. Within this subsection, *C. litoralis* is readily distinguished from *C. rumphii* and related species occurring to the south and east by the distinct stout spines on the male cone sporophylls. *C. thouarsii* R. Br. ex Gaudich. from Indian Ocean localities also has stout spines on the microsporophylls but has narrower leaflets that are highly glaucous when developing. *C. zeylanica* (J. Schust.) A. Lindstr. & K. D. Hill from Sri Lanka and the Andaman islands has longer, more abundant and distinctly sharp cataphylls covered with darker reddish-brown hairs. *C. edentata* de Laub. from the Philippines has particularly glossy leaves, more pointed microsporophylls and smaller megasporophyll blades without teeth.

CONSERVATION STATUS

Cycas litoralis was abundant and widespread in south-east Asia until fairly recently. The plants occur in coastal areas that have experienced rapid growth and development, and numbers have been reducing. The present population number for all countries in which the species is found is estimated to exceed 10 000, the area of occupancy is about 1000 km^2, and the current IUCN Red List status assignment is NT (near threatened) (Hill et al. 2003). This may require reassessment in the future.

HYBRIDS AND INTERGRADES

None known.

USEFUL LITERATURE

Hill et al. (2004) pp. 186–188
Jones (2002) pp. 175–176
Tang et al. (1997) pp. 25–26
Whitelock (2002) pp. 120–121.

FIGURE 47 (d). *Cycas litoralis*. Male cone, Ao Manao Naol, Thailand. Photo: Ken Hill.

FIGURE 47 (e). *Cycas litoralis*. Cluster of megasporophylls, Ao Manao Naol, Thailand. Photo: Ken Hill.

APPENDIX 1: VIETNAMESE PLACE NAMES

We have used the English names for all villages, cities, districts, provinces, nature reserves, etc. in this text. However, we recognise that many maps give place names in the Vietnamese form, and that Vietnamese readers may be more familiar with place names in their own language. The Table below lists the English and Vietnamese equivalents for all proper names used in this volume, except for those where the words are identical.

ENGLISH NAME	VIETNAMESE NAME
Ba Be (National Park, Bac Kan Province)	Ba Bể
Ba Ngoi (Town, Khanh Hoa Province)	Ba Ngòi
Ba Ria–Vung Tau (Province)	Bà Rịa–Vũng Tàu
Bac Kan (=Bac Can, City and Province)	Bắc Kạn
Bac Lieu (City and Province)	Bạc Liêu
Bac Yen (Town, Son La Province)	Bắc Yên
Bach Ma (National Park, Thua Thien–Hue Province)	Bạch Mã
Be Nuoc (Village, Cao Bang Province)	Bể Nước
Ben En (National Park, Thanh Hoa Province)	Bến En
Ben Hai (River, Quang Tri Province)	Bến Hải
Binh Chau-Phuoc Buu (Nature Reserve, Ba Ria–Vung Tau Province)	Bình Châu-Phước Bửu
Binh Dinh (Province)	Bình Định
Binh Thuan (Province)	Bình Thuận
Bu Gia Map (National Park, Binh Phuoc Province)	Bù Gia Mập
Ca Mau (City and Province)	Cà Mâu
Ca Na (Town, Ninh Thuan Province)	Cà Ná
Cam Lap (Village, Khanh Hoa Province)	Cam Lập
Cam Pha (Town, Quang Ninh Province)	Cẩm Phả
Cam Ranh (Town, Khanh Hoa Province)	Cam Ranh
Can Tho (City)	Cần Thơ
Cao Bang (City and Province)	Cao Bằng
Cat Ba (National Park, Hai Phong City)	Cát Bà
Cat Tien (National Park, Dong Nai Province)	Cát Tiên
Cau Hai (Village, Thua Thien–Hue Province)	Cầu Hai
Chi Cong (Town, Binh Thuan Province)	Chí Công
Chi Ne (Town, Hoa Binh Province)	Chi Nê
Chua Huong Tich (Nature Reserve)	Chùa Hương
Con Dao (Islands and National Park, Ba Ria–Vung Tau Province)	Côn Đảo
Con Son (Major island of the Con Dao Archipelago)	Côn Sơn
Cu Mong (Pass, Binh Dinh Province)	Cù Mông
Cuc Phuong (National Park, Ninh Binh Province)	Cúc Phương
Da Krong (Nature Reserve and River, Quang Tri Province)	Đa Krông
Da Nang (City)	Đà Nẵng
Da Trang (Village, Quang Ninh Province)	Đá Trắng
Dak Lak (= Dac Lac, Province)	Đắk Lắk
Deo Ngoan Muc (Nature Reserve, Lam Dong Province)	Đèo Ngoạn Mục
Dien Bien Phu (City, Dien Bien Province)	Điện Biên Phủ
Doc Bao Da (Village, Binh Thuan Province)	Dốc Bảo Đa

Dong Dai (Village, Tuyen Quang Province)	Dông Dài
Dong Dang (Town, Lang Son Province)	Đồng Đăng
Dong Hy (District, Thai Nguyen Province)	Đồng Hỷ
Dong Nai (Province)	Đồng Nai
Fan Si Pan (Mountain, Lao Cai Province)	Phăng Xi Păng (=Hoàng Liên Sơn)
Ha Giang (City and Province)	Hà Giang
Ha Long (City and Bay, Quang Ninh Province)	Hạ Long
Ha Nam (Province)	Hà Nam
Ha Noi (City)	Hà Nội
Ha Tay (Province)	Hà Tây
Ha Tinh (City and Province)	Hà Tĩnh
Hai Phong (City)	Hải Phòng
Hai Van (Pass, Da Nang Province)	Hải Vân
Hat Lot (Town, Son La Province)	Hát Lót
Hin Dan (Mountain, Bac Kan Province)	Hin Đan
Ho Chi Minh City (=Saigon)	Hồ Chí Minh (=Sài Gòn)
Ho Lac (Nature Reserve, Dak Lak Province)	Hồ Lắc
Hoa Binh (Province)	Hòa Bình
Hong (River, flowing through many provinces)	Sông Hồng
Hue (City, Thua Thien–Hue Province)	Huế
Huong Pagoda (Perfume Pagoda, Ha Tay Province)	Chùa Hương
Huu Lien (Nature Reserve, Lang Son Province)	Hữu Liên
Huu Lung (District, Lang Son Province)	Hữu Lũng
Khanh Hoa (Province)	Khánh Hòa
Kien Giang (Province)	Kiên Giang
Lac Thuy (District, Hoa Binh Province)	Lạc Thủy
Lai Chau (City and Province)	Lai Châu
Lam Dong (Province)	Lâm Đồng
Lang Son (City and Province)	Lạng Sơn
Lao Cai (City and Province)	Lào Cai
Lung Trang (Village, Cao Bang Province)	Lũng Trang
Mai Son (District, Son La Province)	Mai Sơn
Minh Hai (Province, now divided into Ca Mau and Bac Lieu Provinces)	Minh Hải
Moc Chau (Town, Son La Province)	Mộc Châu
Mom Ray (Nature Reserve, Kon Tum Province)	Chư Mom Ray
Mui Bai Bung (Peninsula, Ca Mau Province)	Mũi Bãi Bùng
Nghe An (Province)	Nghệ An
Nghia Dan (District, Nghe An Province)	Nghĩa Đàn
Nghia Tan (Village, Ha Giang Province)	Nghĩa Tân
Nha Trang (City, Khanh Hoa Province)	Nha Trang
Ninh Binh (Province)	Ninh Bình
Ninh Thuan (Province)	Ninh Thuận
Pa Nam Cum (Town, Lai Chau Province)	Pa Nậm Cúm
Phang (Pass, Son La Province)	Phặng
Phong Tho (District, Lai Chau Province)	Phong Thổ
Phu Ninh (District, Phu Tho Province)	Phù Ninh

Phu Quoc (Island and National Park, Kien Giang Province)	Phú Quốc
Phu Tho (City and Province)	Phú Thọ
Phu Yen (Province)	Phú Yên
Quang Binh (Province)	Quảng Bình
Quang Nam (Province)	Quảng Nam
Quang Ngai (City and Province)	Quảng Ngãi
Quang Ninh (Province)	Quảng Ninh
Quang Son (Village, Thai Nguyen Province)	Quảng Sơn
Quang Tri (City and Province)	Quảng Trị
Rao An (River, Ha Tinh Province)	Rào Àn
Rao Quan (River, Quang Tri Province)	Rào Quán
Sa Huynh (Town, Quang Ngai Province)	Sa Huỳnh
Son Duong (Town, Tuyen Quang Province)	Sơn Dương
Son La (City and Province)	Sơn La
Son Tra (Nature Reserve, Da Nang City)	Sơn Trà
Song Cau (District, Phu Yen Province)	Sông Cầu
Song Da (River, flowing through many provinces)	Sông Đà
Song Hong (River, flowing through many provinces)	Sông Hồng
Tam Coc (Caves, Ninh Binh Province)	Tam Cốc
Tay Ninh (Province)	Tây Ninh
Thai Nguyen (City and Province)	Thái Nguyên
Thanh Hoa (City and Province)	Thanh Hóa
Thuan Chau (Town, Son La Province)	Thuận Châu
Thuong Tien (Nature Reserve, Hoa Binh Province)	Thượng Tiến
Tram Than (Town, Phu Tho Province)	Trạm Thản
Tuc Nga (Village, Cao Bang Province)	Túc Nga
Tuyen Quang (City and Province)	Tuyên Quang
Vietnam (Country)	Việt Nam
Vinh Hao (Village, Binh Thuan Province)	Vĩnh Hảo
Vinh Phuc (Province)	Vĩnh Phúc
Vo Nhai (District, Thai Nguyen Province)	Võ Nhai
Yen Bai (Province)	Yên Bái
Yen Binh (District, Yen Bai Province)	Yên Bình
Yen Chau, (District, Son La Province)	Yên Châu
Yen Son (District, Tuyen Quang Province)	Yên Sơn
Yok Don (National Park, Dak Lak Province)	Yok Đôn

APPENDIX 2: HERBARIUM ACRONYMS

The list below gives the names of all institutions for which herbarium acronyms have been used in this text. Further address and contact details can be obtained from the website:

http//:sciweb.nybg.org/science2/IndexHerbariorum/asp.

ACRONYM	INSTITUTION
A	Arnold Arboretum of Harvard University, Cambridge, Massachusetts, USA
B	Botanischer Garten und Botanisches Museun Berlin-Dahlem, Zentraleinrichtung der Freien Univerisät Berlin, Berlin, Germany
BKF	Royal Forest Department, Bangkok, Thailand
BM	The Natural History Museum, London, England
FTG	Fairchild Tropical Botanic Garden, Miami, Florida, USA
G	Conservatoire et Jardin Botaniques de la Ville Genèva, Switzerland
G-DC	The de Candolle collection at G
GZAC	Guizhou Agricultural College, Guiyang, Guizhou, China
HITBC	Xishuangbanna Tropical Botanical Garden, Academia Sinica, Xishuangbanna, Yunnan, China
HN	National Center for Natural Sciences and Technology, Hanoi, Vietnam
HNU	Vietnam National University, Hanoi, Vietnam
IBK	Guangxi Institute of Botany, Guilin, Guangxi Zhuangzu, China
K	Royal Botanic Gardens, Kew, England
L	Nationaal Herbarium Nederland, Leiden University Branch, Leiden, Netherlands
LE	V.L. Komarov Botanical Institute, Saint Petersburg, Russia
MO	Missouri Botanical Garden, Saint Louis, Missouri, USA
NF	Nanjing Forestry University, Nanjing, Jiangsu, China
NSW	Royal Botanic Gardens, Sydney, New South Wales, Australia
NY	New York Botanical Garden, Bronx, New York, USA
P	Muséum National d'Histoire Naturelle, Paris, France
PE	Institute of Botany, Chinese Academy of Sciences, Beijing, China
*SFEDI	Sichuan Forestry Exploration and Design Institute, Sichuan, China (*unofficial acronym)
SZG	Shenzhen Fairy Lake Botanical Garden, Shenzhen, Gaungdong, China
U	Nationaal Herbarium Nederland, Utrecht University Branch, Utrecht, Netherlands
XIN	Southwestern Guizhou Institute of Forestry, Xingyi, Guizhou, China

APPENDIX 3: CYCAD-ORIENTED WEBSITES & SOCIETIES

CYCAD E-GROUP

An internet community of over 900 international subscribers. Members learn about cycads, ask questions, provide advice, offer plants for sale or exchange, share photographs, etc.
Website: **http://groups.yahoo.com/group/cycad**

THE CYCAD SOCIETY

Contact: Jody Haynes (Secretary), PO Box 971063, Miami, Florida 33197-1063, USA, tel 305-253-8885, email jody@plantapalm.com. Members receive the quarterly color magazine, *The Cycad Newsletter*, and can participate in the society's seedbank and annual members' meeting. Website: **http://www.cycad.org**

THE CYCAD SOCIETY OF SOUTH AFRICA

Contact: Ian Bassingthwaighte (Secretary), PO Box 176, Montana Park, Pretoria, 0159 South Africa, Tel. 0011-27-12-5481152, email ibass@lantic.net. Members receive the quarterly color magazine, *Encephalartos*, and can participate in the various society branch meetings and excursions. Websites: **http://www.cycadsociety.org** or e-group: **http://groups.yahoo.com/group/sacycad**

MONTGOMERY BOTANICAL CENTER, MIAMI, FLORIDA, USA

A tropical botanical institution specializing in palm and cycad research, conservation and educational collections. Contact: Dr. Patrick Griffith (Executive Director), 11910 Old Cutler Road, Miami, Florida, 33156-4242, USA, tel. 305-667-3800 ext 105, fax 305-661-5984, email: grif@montgomerybotanical.org. Website: **http://www.montgomerybotanical.org**

PACSOA-PALM & CYCAD SOCIETIES OF AUSTRALIA LIMITED

PO Box 1134, Milton, Queensland 4064, Australia, fax 07-3255 8031, email pacsoa1@ozemail.com.au. Members receive the quarterly colour magazine, *Palms & Cycads*, and can participate in the societies' seedbank, bookstore, technical information service, meetings, shows and excursions, with branches and affiliate groups. Website: **http://www.pacsoa.org.au**

THE CYCAD PAGES

Edited by Ken Hill and Dennis Stevenson. An online botanical database based on the *World List of Cycads*, with much useful information on each species. Website: **http://plantnet.rbgsyd.gov.au/PlantNet/cycad**

THE CYCAD SOCIETY OF CHINA

Contact: Prof. C. J. Chen, Institute of Botany, Chinese Academy of Sciences, Xiangshan, Beijing 100093, China,Tel. 0011-86-10-62591431 ext. 6086, Fax 0011-86-10-62590256.

IUCN/SSC CYCAD SPECIALIST GROUP

A component of the World Conservation Union's *IUCN Species Survival Commission*. Contact: Dr. John Donaldson (Chair), Kirstenbosch Research Centre, South African National Biodiversity Institute, Private Bag X7, Claremont, 7735 South Africa, email: Donaldson@sanbi.org. The website, **http://www.iucn.org/themes/ssc/sgs/csg**, also provides many useful links to cycad-oriented research institutions, botanical gardens, nature reserves, and community groups.

PACSOF-PALM & CYCAD SOCIETIES OF FLORIDA, INC., USA

A Florida-based umbrella organization providing a network of people interested in palms and cycads, also holding an annual 'Palm Fest'. The society also maintains the *Virtual Cycad Encyclopedia*, edited by Jody Haynes. Website: **http://www.plantapalm.com/vce/vce_index.htm**

THE PALM & CYCAD SOCIETY OF NEW ZEALAND, INC.

PO Box 3871, Auckland, New Zealand. Members receive the quarterly *NZ Palm & Cycad*, and can participate in the seedbank. Websites: **http://www.geocities.com/palmsnz** or e-group: **http://groups.yahoo.com/group/nzpalmcycad**

THE UNIVERSITY OF CALIFORNIA (BERKELEY), LOS ANGELES, USA

This university website provides an introduction to cycads with several useful links: **http://www.ucmp.berkeley.edu/seedplants/cycadophyta/cycads**.

LITERATURE CITED

Please note that the names of Vietnamese authors are sometimes presented with the given names first (e.g. Loc Phan Ke, Hiêp Tien Nguyen) and sometimes with the surnames first (e.g. Phan Ke Loc, Nguyen Tien Hiêp). To minimise confusion, we have not abbreviated these names in the list of references that follows.

Averyanov, L., P. Cribb, Phan Ke Loc & Nguyen Tien Hiep. 2003. Slipper orchids of Vietnam, with an introduction to the flora of Vietnam. Royal Botanic Gardens, Kew, England.

Brummit, R. K. & C. E. Powell. 1992. Authors of plant names. Royal Botanic Gardens, Kew, England.

Chang, H. T. & Y. C. Zhong. 1995. A floristic analysis of Asian cycads. Chinese Bulletin of Botany (Beijing), Supplement 12: 11–13.

Chang, H. T. & Y. C. Zhong. 1997. New *Cycas* from Guangxi. Acta Scientiarum Naturalium Universitatis Sunyatseni, Guangzhou, China, 36: 67–71. [*Cycas longlinensis*, *C. multifida*, *C. shiwandashanica*, *C. xilingensis*].

Chang, H. T., Y. C. Zhong, Y. Y. Huang, H. X. Zheng & Z. F. Lu. 1998. Additions to the cycadaceous flora of China. Acta Scientiarum Naturalium Universitatis Sunyatseni, Guangzhou, China, 37: 6–87. [*Cycas acuminatissima*, *C. brevipinnata*, *C. longiconifera*, *C. palmatifida*, *C. septemsperma*].

Chen, C. J. 1999. Taxonomical and biogeographical studies on *Cycas* L. (Cycadaceae) in China. Pp. 116–128 in C. J. Chen (ed.), Proceedings of the Fourth International Conference on Cycad Biology. International Academic Publishers, Beijing, China.

Chen, C. J., K. D. Hill & D. W. Stevenson. 2004. Comments on *Cycas*, *Dyerocycas* and *Epicycas* (Cycadaceae). Pp. 57–68 in T. Walters & R. Osborne (eds), Cycad classification—concepts and recommendations. CAB International, Wallingford, England.

Chen, C. J. & D. W. Stevenson. 1999. Cycadaceae. Pp. 1–7 in Z. Y. Wu & P. H. Raven (eds), Flora of China, volume 4, Cycadaceae through Fagaceae. Science Press, Beijing, China and Missouri Botanical Garden Press, Saint Louis, Missouri, USA.

Chen, C. J. & S. Y. Yang. 1994. *Cycas multipinnata* C. J. Chen & S. Y. Yang, a remarkable new cycad from China. Acta Phytotaxonomica Sinica 32: 239, 480–481.

Chen, C. J. & P. Yun. 1995. A new taxa (sic) of *Cycas* (Cycadaceae) in Yunnan. Acta Botanica Yunnanica 17: 400. [*Cycas pectinata* var. *manhaoensis*].

Chen, C. J., L. Zhou, S. Y. Yang & Z. S. Han. 1995. *Cycas* in China, with notes on its conservation status. Pp. 177–191 in P. Vorster (ed.), Proceedings of the Third International Conference on Cycad Biology. The Cycad Society of South Africa, Stellenbosch, South Africa.

Chen, C. J. & Y. Z. Wang. 1995. Taxonomy of *Cycas* L. (Cycadaceae) in China. Chinese Bulletin of Botany (Beijing), supplement 12:14–20.

Cheng, W. C., L. K. Fu & C. Y. Cheng. 1975. Gymnospermae Sinicae. Acta Phytotaxonomica Sinica 13: 81–82.

Craib, W. G. 1912. Contributions to the Flora of Siam II. List of Siamese plants with descriptions of new species. Bulletin of Miscellaneous Information, Kew, pp. 397–435 [*Cycas immersa* sp. nov. , *C. siamensis* Miq.]

Cox, P. A. In print. Biomagnification of cyanobacterial neurotoxins and neurogenerative disease amongst the Chamarro people of Guam. In A. P. Vovides, D. W. Stevenson & R. Osborne (eds), Proceedings of the Seventh International Conference on Cycad Biology.

de Laubenfels, D. J. & F. Adema. 1998. A taxonomic revision of the genus *Cycas* and *Epicycas* gen. nov. (Cycadaceae). Blumea 43: 351–400.

de Loureiro, J. 1793. Pp. 776–777 in Flora Cochinchinensis, volume 2, edition 2. Lisbon, Portugal. [*Cycas inermis*].

Deng, C. Y. 1999. Cycads in Guizhou. Pp. 60–62 in C. J. Chen (ed.), Proceedings of the Fourth International Conference on Cycad Biology. International Academic Publishers, Beijing, China.

Donaldson, J. S. (ed.) 2003. Cycads, status survey and conservation action plan. IUCN/SSC Cycad Specialist Group. IUCN, Gland, Switzerland and Cambridge, England.

Gao, Z. & B. A. Thomas. 1989. A review of cycad megasporophylls with new evidence of *Crossozamia* Pomel and its associated leaves from the Lower Permian of Taiyuan, China. Review of Palaeobotany and Palynology 60: 205–223.

Griffith, W. 1854. Notulae ad Plantas Asiaticas, vol. 4, Bishops College Press, Calcutta, India. [*Cycas dilatata*, *C. jenkinsiana*, *C. pectinata*].

Grobbelaar, N. 2002. Cycads – with special reference

to the southern African species. Published by the author, Pretoria, South Africa.

Guan, Z. T. & G. D. Tao. 1995. Preliminary report on a new cycad species from Yunnan. Sichuan Forestry Survey and Design 4: 1–2, figs 1–7. [*Cycas diannanensis*].

Hall, R. 1998. Reconstructing Cenozoic S E Asia. Pp. 153–184 in R. Hall & J. Holloway (eds), Biogeography and Geological Evolution in SE Asia. Backhuys Publishers, Leiden, Holland.

Hamilton, F. (= F. Buchanan-Hamilton). 1826. Commentary on the Herbarium Amboinense. Memoirs of the Wernerian Natural History Society 5(2): 322–324. [*Cycas pectinata*].

Haynes, J. 2005. Cycad Aulacaspis Scale: a global perspective. The Cycad Newsletter 28(5): 3–6.

Hiêp, Nguyen Tien & R. Kiew. 2000. Thực Vật Tự Nhiên ở Vịnh Hạ Long. Wild Plants of Ha Long Bay. Thanh Nien Publishing House, Vietnam.

Hiêp, Nguyen Tien & Phan Ke Loc. 1997. The cycads of Vietnam (species composition, distribution, ecology, utilisation, threatened status and conservation). Proceedings of the National Centre of Natural Sciences and Technology 9(2): 91–99.

Hiêp, Nguyen Tien & Phan Ke Loc. 1999. The cycads of Vietnam. Pp. 24–32 in C. J. Chen (ed.), Proceedings of the Fourth International Conference on Cycad Biology. International Academic Publishers, Beijing, China.

Hiêp, Nguyen Tien & J. E. Vidal. 1996. Cycadaceae. Pp. 6–23 in Ph. Morat (ed.), Flore du Cambodge, du Laos et du Viêtnam, volume 28, Gymnospermae. Muséum National d'Histoire Naturelle, Paris, France.

Hill, K. D. 1994. The *Cycas rumphii* complex (Cycadaceae) in New Guinea and the western Pacific. Australian Systematic Botany 7: 543–567.

Hill, K. D. 1995. Infrageneric relationships, phylogeny, and biogeography of the genus *Cycas* (Cycadaceae). Pp. 139–162 in P. Vorster (ed.), Proceedings of the Third International Conference on Cycad Biology. The Cycad Society of South Africa, Stellenbosch, South Africa.

Hill, K. D. 2004. Character evolution, species recognition and classification concepts in the Cycadaceae. Pp. 23–44 in T. Walters & R. Osborne (eds), Cycad classification—concepts and recommendations. CAB International, Wallingford, England.

Hill, K. D., C. J. Chen & Phan Ke Loc. 2003. Regional overview: Asia. Pp. 25–30 in Donaldson, J. (ed.) Cycads, status survey and conservation action plan. IUCN/SSC Cycad Specialist Group, IUCN, Gland, Switzerland and Cambridge, England.

Hill, K. D., Hiep Tien Nguyen & Phan Ke Loc. 2004. The genus *Cycas* (Cycadaceae) in Vietnam. Botanical Review 70(2): 134–193.

Hill, K. D. & R. Osborne. 2001. Cycads of Australia. Kangaroo Press, Sydney, Australia.

Hill, K. D. & D. W. Stevenson. 1998–. The Cycad Pages. Website: http://plantnet.rbgsyd.nsw.gov.au/PlantNet/cycad/

Hill, K. D., D. W. Stevenson & R. Osborne. In press. The world list of cycads. In A. P. Vovides, D. W. Stevenson & R. Osborne (eds), Proceedings of the Seventh International Conference on Cycad Biology. Electronic version at: http://www.iucn.org/themes/ssc/sgs/csg/pages/worldlist.htm.

Hill, K. D. & S. L. Yang. 1999a. The cycads of Thailand. Pp. 33–40 in C. J. Chen (ed.), Proceedings of the Fourth International Conference on Cycad Biology. International Academic Publishers, Beijing, China.

Hill, K. D. & S. L. Yang. 1999b. The genus *Cycas* (Cycadaceae) in Thailand. Brittonia 51: 48–73.

Ho, P. H. 1991. Cây-cỏ Việtnam: an illustrated flora of Vietnam. P. H. Ho, Montréal, Canada.

Ho, P. H. & N. Duong. 1960. Cây-cỏ miền nam Việt-nam. Ministry of Education, Saigon, Vietnam.

Huang, S. H., C. J. Chen, Y. M. Shui & Q. R. Li. 1999. A study on the ecology and floristic composition of *Cycas multipinnata* community. Pp. 41–40 in C. J. Chen (ed.), Proceedings of the Fourth International Conference on Cycad Biology. International Academic Publishers, Beijing, China.

IUCN. 1994. IUCN red list categories. IUCN, Gland, Switzerland.

Jones, D. L. 2002. Cycads of the world. Second Edition. Reed New Holland, Sydney, Australia.

Kokubugata, G. & P. I. Forster. 2006. Molecular-cytotaxonomy of *Cycas* (Cycadales) using 5S ribosomal DNA markers. Pp. 163–170 in Y. Tomida et al. (eds), Proceedings of the 7th and 8th Symposia on Collection Building and Natural History Studies in Asia and the Pacific Rim. National Science Museum Monograph 34.

Leandri, J. 1931. Cycadacées. Pp. 1085–1092 in: M. H. Lecomte & F. Gagnepain (eds), Flore Génerále de l'Indo-Chine 5(10). Masson et Cie, Paris, France. [*Cycas chevalieri*, *C. pectinata* var. *elongata*].

Lindström, J. A. 2004. Morphological characters useful in determining species boundaries in *Cycas*

(Cycadaceae). Pp. 45–55 in T. Walters & R. Osborne (eds), Cycad classification—concepts and recommendations. CAB International, Wallingford, England.

Mamay, S. 1976. Paleozoic origin of the cycads. Geological Survey Professional Paper No. 924. U.S. Government, Washington, DC, USA.

Metcalf, F. P. 1942. Flora of Fukien and florisitic notes on southeastern China. Lingnan University, Canton, China.

Miquel, F. A. W. 1861. Prodromus systematis cycadearum. Van der Post, Utrecht, Holland.

Miquel, F. A. W. 1863. [Title unkown]. Botanische Zeitung (Berlin) 21: 334. [*Cycas siamensis*].

Nakai, T. 1943. Ordines, familiae, tribi, genera, sectiones, species, varietates, formae et combinationes novae a Prof. Nakai-Takenosin adhuc ut novis edita, Appendix Quaestiones characterium naturalim plantarum. Universitatis Imperialis Tokyoensis, Tokyo, Japan. [*Dyerocycas micholitzii*, as *D. mischolitzi*].

Nguyen, M. T. 1995. Hệ thống các Vườn quốc gia và Khu bảo tồn thiên nhiên. System of National Parks and Nature Reserves. Pp. 14–19 & 163–167 in Proceedings of the National Conference on National Parks and Protected Areas in Vietnam. Agricultural Publishing House, Hanoi, Vietnam.

Norstog, K. J. & T. J. Nicholls. 1997. The biology of the cycads. Cornell University Press, Ithaca, New York, USA.

Osborne, R. 2002. Cycad fossils. Encephalartos 69: 4–13.

Osborne, R. 2005. Focus on ... *Cycas tropophylla*. Encephalartos 82: 8–13.

Osborne, R. & K. D. Hill, Hiêp Tien Nguyen & Loc Phan Ke. 2006. Focus On... *Cycas hoabinhensis*. Encephalartos 87: 9–13.

Osborne, R. & R. Gorelick. In press. Sex change in cycads—cases, causes and chemistry. In A. P. Vovides, D. W. Stevenson & R. Osborne (eds), Proceedings of the Seventh International Conference on Cycad Biology.

Osborne, R. & Tien Hiep Nguyen. 2002. Focus on ... *Cycas elongata*. Encephalartos 72: 4–9.

Osborne, R. & T. Walters. 2004. Glossary of terms encountered in cycad systematics. Appendix 2, pp. 237–258 in T. Walters & R. Osborne (eds), Cycad classification—concepts and recommendations. CAB International, Wallingford, England. Electronic version available at: http://www.iucn.org/themes/ssc/sgs/csg/publications/CCC-Glossary.pdf.

Pant, D. D. 1962. *Cycas* and the cycadales. Central Book Depot, Allahabad, India. First edition 1962, Second edition 1973.

Pant, D. D., R. Singh & D. K. Chauhan. 1994. On *Cycas pectinata* Hamilton from north-east India. Encephalartos 38: 17-30.

Phan Ke Loc.1998. On the systematic structure of the Vietnamese flora, pp. 120–129 in A. Zhang and S. Wu (eds), Floristic Characteristics and Diversity of East Asian Plants. Springer Verlag and China Higher Education Press, Beijing, China, and Berlin, Germany.

Phan Kế Lộc, Nguyễn Tiến Hiệp, Nguyễn Thị Thanh Hương, Đỗ Tiến Đoàn, Nguyễn Quang Hiếu, Tô Văn Thảo, Nguyễn Tiến Vinh, Nguyễn Sinh Khang, Phạm Văn Thế, Dương Thị Hoàn, J. Regalado Jr., L. V. Averyanov, A. L. Averyanova, D. D. Soejarto, K. D. Hill, S. K. Wu, J. Wen, A. Farjon, S. Swan, N. Furey, Nguyễn Mạnh Cường & Vũ Ngọc Long. 2005. Một số taxôn mới hay/và bổ sung cho hệ thực vật Việt Nam và vấn đề bảo tồn chúng. Trong Những vấn đề nghiên cứu cơ bản trong khoa học sự sống. Báo cáo khoa học, Hội nghị toàn quốc 2005. Nghiên cứu cơ bản trong khoa học sự sống. Đại học Y Hà Nội 03-11-2005. Nhà xuất bản Khoa học và Kỹ thuật. Hà Nội: 15–19. [Some new discoveries to science and/or for the flora of Vietnam and problems of their conservation, pp. 15–19 in: Issues of basic research in life sciences. Proceedings of the National Conference on Life Sciences, Hanoi University, 3 November 2005. Science and Technics Publishing House, Hanoi.]

Prain, D. 1909. *Cycas micholitzii*. Curtis Botanical Magazine 135: t.8242 & text.

Raffill, C. P. 1912. *Cycas micholitzii*. Bulletin of miscellaneous information, Kew, pp. 301–302.

Red Data Book of Vietnam. 1996. Bộ Khoa học, Công nghệ và Môi trường. Sách đỏ Việt Nam–Phần Thực vật. Nhà xuất bản Khoa học và Kỹ thuật. Hà Nội. [Red Data Book of Vietnam. Volume 2. Plants. Ministry of Science, Technology and Environment. Science and Technics Publishing House. Hanoi].

Regalado, J., D. Harder, Nguyen Tien Hiep, Nguyen Thi Thanh Huong, L. Averyanov and Phan Ke Loc. 2003a. Main results of international cooperative research in the study of plant diversity in Vietnam, 1993–2002. Vietnam National University Journal of Science, Natural Science & Technology XIX, 4:

63–72.

Regalado, J., D. Harder, Nguyễn Tiến Hiệp, Nguyễn Thị Thanh Hương, L. Averyanov and Phan Kế Lộc. 2003b. Các taxôn thực vật bậc cao có mạch mới cho khoa học và/hoặc bổ sung cho hệ thực vật Việt Nam. Trong Những vấn đề nghiên cứu cơ bản trong khoa học sự sống. Nhà xuất bản Khoa học và Kỹ thuật. Hà Nội, các trang 145–149.

Regalado, J., Nguyen Tien Hiep, Phan Ke Loc, L. Averyanov & D. K. Harder. 2005. New insights into the diversity of the Flora of Vietnam. Biologiske Skrifter 55: 189–197.

Sacks, O. 1996. The Island of the colour-blind. Pan Macmillan, Sydney, Australia.

Schuster, J. 1932. Cycadaceae. Pp.1–168 in A. Engler (ed.), Das Pflanzenreich 99(IV.I). W. Engelmann, Leipzig, Germany.

Shaw, C. A., J. M. B. Wilson, R. Cruz-Aguado, S. Singh, E. L. Hawkes, V. Lee and T. Marler. In print. Cycad-induced degeneration in a mouse model of ALS-PDC: is the culprit really BMAA or is a novel toxin to blame? In A. P. Vovides, D. W. Stevenson & R. Osborne (eds), Proceedings of the Seventh International Conference on Cycad Biology.

Smitinand, T. 1971. The genus *Cycas* Linn. (Cycadaceae) in Thailand. Natural History Bulletin of the Siam Society 24(1–2): 163–175. [*Cycas micholitzii* var. *simplicipinna*].

Smitinand, T. 1972. Cycadaceae, in T. Smitinand & K. Larsen (eds), Flora of Thailand 2(2): 185–192.

Stearn, W. T. 1992. Botanical Latin. Fourth edition. David & Charles, Devon, England.

Sterling, E. J., M. M. Hurley and Le Duc Minh. 2006. An Introduction to the natural history of Vietnam. Yale University Press, USA.

Suvatabandhu, K. 1961. The living gymnosperms of Thailand. Journal of the National Research Council of Thailand 2: 59–62.

Tang, W. 2003. Focus on … *Cycas lindstromii*. Encephalartos 76: 8–12.

Tang, W. 2004. Continental drift and the evolution of Asian *Cycas*. Encephalartos 80: 23–28.

Tang, W., W. L. Chiu & H. Pu. 2001. The bamboo *Cycas* and its relatives: taxonomy and distribution. Palms & Cycads 72: 3–11.

Tang, W., J. A. Lindström & L. Nian. 2004. *Cycas debaoensis* Conservation Project in China. Pp. 77–83 in J. A. Lindström (ed.), Proceedings of the Sixth International Conference on Cycad Biology. Nong Nooch Tropical Botanical Garden, Chonburi, Thailand.

Tang, W., R. G. Oberprieler & S. L. Yang. 1999. Beetles (Coleoptera) in cones of Asian *Cycas*: diversity, evolutionary patterns, and implications for *Cycas* taxonomy. Pp. 280–297 in C. J. Chen (ed.), Proceedings of the Fourth International Conference on Cycad Biology. International Academic Publishers, Beijing, China.

Tang, W. & S. L. Yang. 1995. Focus on … *Cycas siamensis*. Encephalartos 43: 4–11.

Tang, W., S. L. Yang & P. Vatcharakorn. 1997. Cycads of Thailand. Nong Nooch Tropical Botanical Garden, Chonburi, Thailand.

Thiselton-Dyer, W. T. 1902. Cycadaceae, in F. B. Forbes & W. B. Hemsley's enumeration of the known plants from China proper, Formosa, etc. Journal of the Linnean Society, Botany 26: 559–561. [*Cycas rumphii* var. *bifida*].

Thiselton-Dyer, W. T. 1905. *Cycas micholitzii*. New or noteworthy plants. Gardeners' Chronicle, series 3, 38(973): 142–144.

Vietnamese Government Decree. 2002. No. 48/2002/ND-CP, dated 22/4/2002 on 'Amending and Supplementing the List of Rare and Precious species of Flora and Fauna and Regulations for their Management and Protection' and Official Dispatch No. 3399/VPCP-NN of the Council of Ministers, dated 21/6/2002, on Amending the list of Rare and Precious Species of Flora and Fauna Propagated in Decree 48.

Vietnamese Government Decree. 2006. Nghị định số 32/2006/NĐ–CP của Chính phủ ngày 30 tháng 3 năm 2006 Về quản lý thực vật rừng, động vật rừng nguy cấp, quý, hiếm. [Decree No. 32/2006/ND–CP, dated 30/3/2006, on the Management of Threatened, Rare and Precious species of Flora and Fauna of Vietnam.]

Vietnamese Government publication. 2003. Cộng hòa xã hội chủ nghĩa Việt Nam: Chiến lược quản lý hệ thống Khu bảo tồn thiên nhiên Việt Nam đến năm 2010. Hà Nội. [Management strategy for a protected areas system in Vietnam to 2010.]

Vorster, P. 1996. Encounters with Chinese cycads. Encephalartos 48: 15-24.

Vorster, P. 2002. The Vietnam cycad expedition. Part 1, Encephalartos 72: 16–21.

Vorster, P. 2003a. The Vietnam cycad expedition. Part 2, Encephalartos 73: 17–24.

Vorster, P. 2003b. The Vietnam cycad expedition. Part 3, Encephalartos 74: 34–39.

Walters, T. W., S. L. Yang, H. Pu & D. S. Decker-Walters. 1995. Cycads of China: observations from a 1992 expedition. Pp. 163–176 in P. Vorster (ed.), Proceedings of the Third International Conference on Cycad Biology. The Cycad Society of South Africa, Stellenbosch, South Africa.

Wang, D. Y. 1995. A preliminary study of the *Cycas micholitzii* complex. Encephalartos 44: 31–38.

Wang, D. Y. 1996. Taxonomy of *Cycas* in China. Pp. 33–142 in F. X. Wang & H. B. Laing (eds), Cycads in China. Guangdong Science and Technology Press, Guangdong, China.

Wang, D. Y. & C. Y. Deng. 1995. *Cycas segmentifida*, a new species from China. Encephalartos 43: 11–14.

Warburg, O. 1900. Species of southern Asia. Cycadaceae. Monsunia: 178–181. [*Cycas balansae*, *C. miquellii*].

Wei, F. N. 1994. A new cycad from Guangxi. Guihaia 14: 300. [*Cycas ferruginea*].

Wei, F. N. 1996. Studies on the resources and classification of the wild cycads from Guangxi 1. Guihaia 16: 1–2. [*Cycas sexseminifera*].

Wei, F. N., J. Y. Liang & S. Q. He. 1997. Studies on the resources and classification of the wild cycads from Guangxi 2. Guihaia 17: 206–212. [*Cycas spiniformis*, *C. longisporophylla*].

Whitelock, L. M. 2002. The cycads. Timber Press, Portland, Oregon, USA.

Xiao, L. Q. & X. Gong. 2006. Genetic differentiation and relationships of populations in the *Cycas balansae* complex (Cycadaceae) and its conservation implications. Annals of Botany 97: 807–812.

Yang, S. L., K. D. Hill & Nguyen Tien Hiêp. 1997. *Cycas lindstromii* (Cycadaceae), a new species from southern Vietnam. Novon 7: 213–215.

Zhou, L., S. Y. Yang & Z. Zhou. 1990. Investigation of the natural community of *Cycas panzhihuaensis* L. Zhou & S. Y. Yang. Pp. 148–151 in D. W. Stevenson (ed.), Proceedings of the First International Conference on Cycad Biology. Memoirs of the New York Botanical Garden 57.

GLOSSARY OF TERMS

acute. Ending in a sharp point.

adventitious. Vegetative growth on a cycad stem leading to suckers at the base or offsets on the stem.

apex. Tip of an organ.

arborescent. Tree-like in nature.

bipinnate. Twice pinnate; a compound leaf having first and second order divisions.

bract. Leaf-like structure subtending a shoot or axillary bud, usually having a protective function. Sometimes misapplied to cataphylls of cycads.

cataphyll. A modified leaf, reduced and thickened, protecting the apical meristem of cycads.

caudex. Thick stem or trunk.

character. An identifiable and hereditable feature, useful when comparing one species with another.

circinate. Rolled up in a coil-like fashion, like the pinnae of emerging leaves of *Cycas* species.

cladogram. Tree diagram based on parsimony analysis. Taxa are shown grouped hierarchically into nested sets according to their synapomorphies. It implies no timeline and shows no ancestral connotation.

cone. The reproductive structure of gymnosperms; collection of sporophylls organised on a central axis.

coralloid root. Apogravitropic, club-shaped root, potentially hosting cyanobacteria.

cyanobacteria. Group of bacteria which can photosynthesize, found in cycad roots.

Cycadales. Gymnosperm plants order, containing all extant and extinct cycads.

cycads. Members of the order Cycadales.

decumbent. Stems lying along the ground but turned up at the apex.

dichotomous. Branching into two equal parts like the stems of *Cycas elongata*.

dioecious. Male and female reproductive parts occur on different plants.

disjunct. Geographically separated, as in populations of the same species occurring in more than one geographical area.

DNA. Deoxyribonucleic acid, the molecule in which an organism's definition is coded.

ecology. The study of relationships of living organisms and their environment, physical and chemical.

embryo. In seeds the diploid tissue arising from the zygote, until germination.

endemic. Occurring in a restricted area.

epithet. Second part of the binomial, the specific epithet.
etymology. Dealing with the origin of words. Species descriptions give the etymology for the specific epithet.
extant. Existing currently. Opposite of extinct.
falcate. Sickle-shaped; curved.
glabrous. Of a smooth surface without any hair.
glaucous. Of a surface with a bluish grey waxy or powdery bloom.
isotype. Duplicate herbarium specimen of the holotype.
lamina. Flattened, expanded part of a leaf, sporophyll or similar structure.
leaflet. Pinna; primary division or a compound leaf
lectotype. Herbarium specimen chosen from the original material to replace a holotype.
megasporophyll. Sporophyll bearing one or more ovules or later seeds.
microsporangium. Structure on the bottom surface of the microsporophyll; pollen sacs.
microsporophyll. Sporophyll bearing microsporangia.
monophyletic. Derived from a single ancestor.
neotype. New material to replace a missing holotype.
pachycaulous. Thick-stemmed and columnar but without any substantial secondary wood.
pectinate. Comb-like; closely spaced, narrow segments.
petiole. Stalk of a leaf.
r(h)achis. Part of axis of compound leaf where pinnae (leaflets) are attached, excluding the petiole.
recurved. Curving abaxially.
revolute. With margins rolled downward.
sarcotesta. Fleshy tissue layer of the integument of cycad seed, often brightly coloured.
sclerotesta. Hard, stony inner layer of integument of cycad seed.
speciation. Evolutionary process through which plants accumulate enough differences to be recognised as a species.
species. Basic taxonomic rank, below genus.
spinescent. Ending in a sharp point.
synapomorphy. Shared derived character state that unites two or more members of a monophyletic group.
taxon. A group of individuals, in any rank, sharing common characteristics and of the same evolutionary origin.
tomentose. Densely woolly in a finely matted fashion.
tomentum. Covering of fine hair.
type. Herbarium specimen to which the name of a species is permanently attached, whether it is a correct name or synonym.
undulate. Having wavy margins.

LIST OF SUBSCRIBERS

Greg Baetge	USA
Louis Biagi	USA
Alison Botha	South Africa
Wayne Boyd	Australia
Sathitvong Chayangsu	Thailand
Jeffrey Cheffet	USA
Jeff Chemnick	USA
Gary Coleman	New Zealand
Linda Costa	Australia
Carey Delzell	USA
Root Gorelick	Canada
Adam Grainger	Australia
Tim Gregory	USA
Nat Grobbelaar	South Africa
Don Hemmes	USA
Royal Botanic Gardens, Kew	England
John Lok	New Zealand
Todd Maerowitz	USA
Irv McDaniel	USA
Montgomery Botanical Center	USA
Craig Nazor	USA
Rolf Oberprieler	Australia
Harvey Ottley	Australia
Joe Perner	Australia
Michael Perry	USA
Marion Peter	Germany
Werner Schwabe	Germany
Hideo Shimizu	Japan
Rita Singh	India
William Tang	USA
Eric Thoumire	France
Loran Whitelock	USA
Glen Williams	USA
Enno Worbs	Germany
Neil Yorio	USA